Ernst Schering Foundation Symposium
Proceedings 2006-2
GPCRs: From Deorphanization
to Lead Structure Identification

Ernst Schering Foundation Symposium
Proceedings 2006-2

GPCRs: From Deorphanization to Lead Structure Identification

H. Bourne, R. Horuk, J. Kuhnke, H. Michel
Editors

With 59 Figures

Springer

Series Editors: G. Stock and M. Lessl

Library of Congress Control Number: 2007921597

ISSN 0947-6075

ISBN 978-3-540-48981-8 Springer Berlin Heidelberg New York

Springer is a part of Springer Science+Business Media
springer.com

Editor: Dr. Ute Heilmann, Heidelberg
Desk Editor: Wilma McHugh, Heidelberg
Production Editor: Anne Strohbach, Leipzig
Cover design: WMXDesign, Heidelberg
Typesetting and production: LE-TEX Jelonek, Schmidt & Vöckler GbR, Leipzig
21/3180/YL – 5 4 3 2 1 0 Printed on acid-free paper

Preface

G-protein-coupled receptors (GPCRs) play a major role in regulating the overall homeostasis of complex organisms such as mammals but are also found in primitive species such as Dictyostelium (slime mold) and yeast. The GPCR superfamily is quite diverse and sequencing has revealed more than 850 genes comprising approximately 3% of the human genome. The diversity of the GPCRs is equally matched by the variety of ligands that activate them, which include odorants, taste ligands, light, metals, biogenic amines, fatty acids, amino acids, peptides, proteins, nucleotides, lipids, Krebs cycle intermediates, and steroids. Because of their central role in regulating normal physiological responses, GPCRs have attracted considerable attention from the pharmaceutical industry as targets for disease. This large superfamily of proteins remains one of the most druggable targets, accounting for more than 40% of all marketed therapeutics.

Based on their ever-growing importance, as outlined above, a recent Ernst Schering Research Foundation Workshop held in Berlin, Germany, focused on GPCRs. Entitled "GPCRs: From Deorphanization to Lead Structure Identification," the workshop brought together leading experts from a variety of areas to discuss recent advances in the field. Professor Henry Bourne of UCSF gave an enthralling keynote lecture entitled "G-Proteins and GPCRs: From the Beginning" This lecture chronicled the progress made from Earl Sutherland's original Nobel

Prize-winning achievement in discovering cyclic AMP as a second messenger molecule to the discovery of G-proteins and Gapers. As Professor Bourne reminded us, these breakthroughs were all the more remarkable because they were achieved in the absence of many of the breakthrough technologies we all take for granted nowadays such as molecular cloning and PCR.

The formal presentations on the second day began with a lecture from Professor Hartman Michel of the Max Planck Institute in Frankfurt, Germany, who discussed approaches to obtain protein crystals of GPCRs. As Professor Michel reminded us, despite massive efforts only one GPCR, bovine rhodopsin, has so far succumbed to this approach. Professor Michel's presentation centered on the methods that he and other groups are using with expression systems to try to overcome some of the hurdles in obtaining sufficient active protein to enable crystallization. At this point, the workshop shifted gears and switched from discussing models determined from a protein crystal of a GPCR to models derived from analysis of GPCRs by in silico methods. First up was Professor Gert Vriend of the CMBI at Nijmegen with a thought-provoking presentation centered on molecular modeling of GPCRs. Comparing predictions of GPCR structure, Professor Vriend noted that numerous problems and errors remain inherent in most of these models, even those based on the crystal structure of bovine rhodopsin. He pointed out that although homology modeling of GPCRs has reached a vogue in drug discovery, it needs to be approached with some caution and is not quite as predictable as its proponents would have us believe. Following this controversial viewpoint, Professor Alex Tropsha of UNC Chapel Hill continued in a similar vein, albeit more optimistically, with a presentation that examined the role of QSAR in aiding and abetting drug discovery for GPCRs. Here the idea is that known small molecule ligands for a given GPCR can be used to build models with which to interrogate the chemical universe to discover new structures for that receptor. Professor Tropsha pointed out that ligand-based modeling has been successfully applied to a number of receptors, including those from the dopaminergic and serotinergic receptor families, which he used as examples.

The well-established idea that particular classes of ligand substructures seem to occur quite frequently in pharmacologically successful small molecules was the theme developed by Dr. Robert Bywater of Magdalen College, Oxford. The notion of privileged structures in GPCRs has been exploited in drug design and, as pointed out by Dr. Bywater, has been useful to generate both receptor agonists and antagonists. He concluded his seminar by suggesting that the concept of privileged structures might be useful in designing drugs for orphan GPCRs whose natural ligands have not yet been identified. Two further speakers highlighted the treasure trove of targets that have yet to be discovered from the large pool of orphan GPCRs. Dr. Alan Wise of GSK, Harlow, UK, gave several examples of receptor deorphanization approaches used at GSK. In particular he pointed out the use of a knowledge-based approach for the successful pairing of nicotinic acid with the receptor HM74. Nicotinic acid has been used clinically for over 40 years to treat dyslipidemia acting on adipose cells. Expression analysis revealed that ten orphan GPCRs were expressed in adipose tissue. These were recombinantly expressed and binding experiments revealed that HM74 was the nicotinic acid receptor. Professor Marc Parmentier from the University of Brussels, Belgium, continued along the same lines, highlighting the successful strategies for deorphanizing GPCRs currently employed in his group. Reverse pharmacological approaches that involve extraction of putative ligands for GPCRs from tissue extracts then testing them for activity on orphan GPCRs were a feature of his presentation. An example of this was the identification of nocipeptin as the ligand for ORL1 and apelin as the ligand for the APJ receptor. Dr. Parmentier finished his seminar by pointing out that some orphan GPCRs might only exist as receptor heterodimers, which would be incredibly difficult to deorphanize. The concept of GPCR dimers was amply illustrated from his own work with chemokine receptors, for example CCR2/CCR5, which he showed can exist as heterodimers.

Clearly, the existence of chemokine receptors as dimers will have a profound effect on drug discovery since many existing paradigms and concepts will have to be altered if we are to be successful in finding drugs that target these complexes. Professor Graeme Milligan of the University of Glasgow, Scotland, UK, took us down this new avenue with his presentation looking at the role of GPCR dimerization in receptor signaling. Professor Milligan reminded us that the only established example of GPCR dimers comes from atomic force microscopy of murine rod outer-segment discs that reveal that rhodopsin is organized in a series of parallel arrays of dimers. This organization of GPCRs as homo- or heterodimers may be more common than we imagine and Professor Milligan illustrated this with examples such as the alpha adrenergic receptor and the interesting heterodimer formed between the cannabinoid 1 and the orexin receptors, which explain the pharmacological action of the appetite suppressant Rimonabant.

Dr. Rob Leurs of the Vrije Universiteit Amsterdam, The Netherlands, reminded us that viruses have exploited human GPCRs very effectively to overcome host defense mechanisms so that they can propagate. A chilling example is the use of the chemokine receptors CCR5 and CXCR4 as vehicles of entry for HIV-1, which gives rise to the deadly disease AIDS. Dr. Leurs showed that viruses could also express GPCRs, potentially pirated from their hosts. An example of this was the human cytomegalovirus virus US28, which has been associated with chronic diseases and malignancies. Examples of targeting these and other viruses with small molecule antagonists might open up new avenues of treatment for some human diseases.

Professor Eric Prossnitz of the University of New Mexico, Albuquerque, returned us right back to Henry Bourne's opening address with his discussion of receptor signaling. His theme was that GPCRs exist in a large variety of conformations. These can be ligand-induced or can be induced by post-translational modifications. As Professor Prossnitz reminded us, these conformations are unique and might represent novel targets for drug discovery and therapeutic intervention.

Two excellent presentations by representatives from the pharmaceutical industry highlighted approaches in drug discovery for GPCRs. The first by Dr. Ralf Heilker from Boehringer Ingelheim, Biberach an der Riss, Germany, explained the advantages of high-content screening to monitor G-protein-coupled receptor internalization as a means of drug discovery. High-content screening is a combination of fluorescence microscopic imaging and automated image analysis, and has found increasing use in monitoring the effects of compounds in cellular systems, for example receptor desensitization, in which receptors internalize after ligand stimulation. The use of such assays to pharmacologically profile compounds was very nicely demonstrated by Dr. Heilker. Dr. Andreas Sewing of Pfizer outlined drug discovery approaches employed in his company to generate therapeutics targeting GPCRs. His seminar centered on high-throughput screening approaches for rapidly discovering lead compounds. The choice of the assay employed is obviously important for success, as already discussed by the previous speaker. There was further discussion on reagent generation and supply and lead-finding strategies applied to biological screening. Finally, the hit-to-lead and lead-optimization processes were discussed.

All in all, the meeting greatly exceeded all of our expectations and lived up to the ideals of the Ernst Schering Research Foundation Workshop to sponsor meetings that bring together a critical mass of top scientists working in important areas in an intimate setting that fosters the free exchange of knowledge and ideas.

H. Bourne
R. Horuk
J. Kuhnke
H. Michel

Contents

List of Editors and Contributors

Editors

Bourne, H.R.
UCSF, 600 16th Street, San Francisco, CA 94143-2140, USA
(e-mail: bourne@cmp.ucsf.edu)

Horuk, R.
Berlex Biosciences, 2600 Hilltop Drive, Richmond, CA 94806, USA
(e-mail: Horuk@pacbell.net)

Kuhnke, J.
Schering AG, Müllerstr. 178, 13342 Berlin, Germany
(e-mail: joachim.kuhnke@bayerhealthcare.com)

Michel, H.
Max-Planck Institut für Biophysik,
Max-von-Laue-Straße 3, 60438 Frankfurt-am-Main, Germany
(e-mail: Hartmut.Michel@mpibp-frankfurt.mpg.de)

Contributors

Bywater, R.P.
Magdalen College, Oxford High Street, Oxford OX1 4AU, UK
(e-mail: robert.bywater@magd.ox.ac.uk)

Canals, M.
Molecular Pharmacology Group,
Division of Biochemistry and Molecular Biology,
Institute of Biomedical and Life Sciences,
University of Glasgow, Glasgow G12 8QQ, UK

Cawkill, D.
Lead Discovery Technology, Pfizer PDGRD, IPC 580, Ramsgate Road,
Sandwich CT13 9NJ, UK

Detheux, M.
Euroscreen S.A, Rue A. Bolland 47, 6041 Gosselies, Belgium

de Esch, I.J.P.
Leiden/Amsterdam Center for Drug Research (LACDR),
Division of Medicinal Chemistry, Faculty of Sciences,
Vrije Universiteit Amsterdam, De Boelelaan 1083,
1081 HV Amsterdam, The Netherlands

Ellis, J.
Molecular Pharmacology Group,
Division of Biochemistry and Molecular Biology,
Institute of Biomedical and Life Sciences,
University of Glasgow, Glasgow G12 8QQ, UK

Heilker, R.
Boehringer Ingelheim Pharma GmbH Co. KG,
Department of Lead Discovery, Birkendorfer Strasse 65,
88397 Biberach an der Riss, Germany
(e-mail: Ralf.Heilker@bc.boehringer-ingelheim.com)

Horn, F.
Laboratoire de Biologie, Informatique et Mathématiques,
CEA, Grenoble, France

Hulshof, J.W.
Leiden/Amsterdam Center for Drug Research (LACDR),
Division of Medicinal Chemistry, Faculty of Sciences,
Vrije Universiteit Amsterdam, De Boelelaan 1083,
1081 HV Amsterdam, The Netherlands

Jacoby, E.
Novartis Institutes for BioMedical Research, Discovery Technologies,
Lichtstrasse 35, 4056 Basel, Switzerland
(e-mail: edgar.jacoby@novartis.com)

Leurs, R.
Leiden/Amsterdam Center for Drug Research (LACDR),
Division of Medicinal Chemistry, Faculty of Sciences,
Vrije Universiteit Amsterdam, De Boelelaan 1083,
1081 HV Amsterdam, The Netherlands
(e-mail: r.leurs@few.vu.nl)

Lopez-Gimenez, J.F.
Molecular Pharmacology Group, Division of Biochemistry
and Molecular Biology, Institute of Biomedical and Life Sciences,
University of Glasgow, Glasgow G12 8QQ, UK

Milligan, G.
Molecular Pharmacology Group, Division of Biochemistry
and Molecular Biology, Institute of Biomedical and Life Sciences,
University of Glasgow, Glasgow G12 8QQ, UK
(e-mail: g.milligan@bio.gla.ac.uk)

Oliveira, L.
Escola Paulista de Medicina, Sao Paulo, Brazil

Paiva, A.C.M.
Escola Paulista de Medicina, Sao Paulo, Brazil

Parmentier, M.
IRIBHN, ULB Campus Erasme,
808 route de Lennik, 1070 Brussels, Belgium
(e-mail: mparment@ulb.ac.be)

Pediani, J.D.
Molecular Pharmacology Group, Division of Biochemistry
and Molecular Biology, Institute of Biomedical and Life Sciences,
University of Glasgow, Glasgow G12 8QQ, UK

Prossnitz, E.R.
Department of Cell Biology and Physiology, Cancer Research
and Treatment Center, University of New Mexico Health
Sciences Center, Albuquerque, NM 87131, USA
(e-mail: eprossnitz@salud.unm.edu)

Sewing, A.
Lead Discovery Technology, Pfizer PDGRD, IPC 580, Ramsgate Road,
Sandwich CT13 9NJ, UK
(e-mail: Andreas.Sewing@pfizer.com)

Sklar, L.A.
Department of Cell Biology and Physiology, Cancer Research
and Treatment Center, University of New Mexico Health
Sciences Center, Albuquerque, NM 87131, USA

Smit, M.J.
Leiden/Amsterdam Center for Drug Research (LACDR),
Division of Medicinal Chemistry, Faculty of Sciences,
Vrije Universiteit Amsterdam, De Boelelaan 1083,
1081 HV Amsterdam, The Netherlands

Tropsha, A.
The Laboratory for Molecular Modeling, CB#7360, Beard Hall,
School of Pharmacy, University of North Carolina at Chapel Hill,
Chapel Hill, NC 27599-7360, USA
(e-mail: alex_tropsha@unc.edu)

Vischer, H.F.
Leiden/Amsterdam Center for Drug Research (LACDR),
Division of Medicinal Chemistry, Faculty of Sciences,
Vrije Universiteit Amsterdam, De Boelelaan 1083,
1081 HV Amsterdam, The Netherlands

Vriend, G.
CMBI NCMLS, UMC, Geert Grooteplein 28, 6525 GA Nijmegen,
The Netherlands
(e-mail: Vriend@CMBI.ru.nl)

Wang, S.X.
The Laboratory for Molecular Modeling, School of Pharmacy,
University of North Carolina at Chapel Hill,
Chapel Hill, NC 27599-7360, USA

Wigglesworth, M.J.
Screening and Compound Profiling, GlaxoSmithKline, New Frontiers
Science Park, Third Avenue, Harlow, Essex, CM19 5AW, UK
(e-mail: Mark.J.Wigglesworth@gsk.com)

Wise, A.
Screening and Compound Profiling, GlaxoSmithKline, New Frontiers
Science Park, Third Avenue, Harlow, Essex, CM19 5AW, UK
(e-mail: Alan.x.Wise@gsk.com)

Wolfe, L.A. III
Screening and Compound Profiling, GlaxoSmithKline, New Frontiers
Science Park, Third Avenue, Harlow, Essex, CM19 5AW, UK
(e-mail: Larry.A.Wolfe@gsk.com)

Ernst Schering Foundation Symposium Proceedings, Vol. 2, pp. 1–21
DOI 10.1007/2789_2006_001

Published Online: 16 May 2007

G-Proteins and GPCRs: From the Beginning

H.R. Bourne(✉)

UCSF, 600 16th Street, 94143-2140 San Francisco, USA
email: *bourne@cmp.ucsf.edu*

Abstract. From the point of view of a participant observer, I tell the discovery stories of trimeric G-proteins and GPCRs, beginning in the 1970s. As in most such stories, formidable obstacles, confusion, and mistakes make eventual triumphs even more exciting. Because these pivotally important signaling molecules were discovered before the recombinant DNA revolution, today's well-trained molecular biologist may find it amazing that we learned anything at all.

Born three decades ago and now grown to robust maturity, trimeric G-proteins and G-protein-coupled receptors (GPCRs) continue to generate exciting advances in biology and drug discovery. Here I recount the story of their births, from the point of view of a participant observer. As in most discovery stories, formidable obstacles, confusion, and mistakes make eventual triumphs even more satisfying.

Two unrelated events—Sutherland's discovery of cAMP in the 1950s and the Vietnam war of the 1960s—brought me into the story. To avoid military service, I spent two years at the National Institutes of Health (NIH), in Bethesda, Maryland, where I learned to measure cAMP synthesis in fat cell extracts. In 1969 I moved to the University of California San Francisco (UCSF) as a research fellow, and began to study cAMP in human leukocytes, a choice that reflected widespread interest in cAMP as a second messenger, plus the fact that no one else west of the Mississippi river knew the adenylyl cyclase assay. Neither I nor my colleagues could have foreseen the delights cAMP would eventually bring.

Indeed, signaling research in the 1960s and 1970s would be almost unrecognizable to scientists trained after the recombinant DNA revolution of the 1980s. The cutting edge was hard-core biochemistry, but many experiments focused on bio-assays using animal tissues or enzyme assays in extracts. In multiple laboratories from 1964 to 1972, I never heard the words "genetics", "DNA", or "evolution" mentioned, much less used in planning an actual experiment. Today's molecular biologists will find it astonishing that we learned anything at all.

I shall tell the birth stories of G-proteins and GPCRs in more or less chronological order, emphasizing what investigators thought and imagined at the time and explicitly labeling explanations based on hindsight. A caveat is in order: more memoir than scholarly treatise, these stories necessarily reflect a personal point of view, replete with limitations of observer bias, faulty memory, and ignorant omission. Nonetheless, the message is as true as I can make it, even if some details are wrong.

1 Prologue: GTP Enters the Picture

In the early 1970s, Martin Rodbell's laboratory at the NIH was assaying adenylyl cyclase and binding of radioactive glucagon in liver membranes. Lutz Birnbaumer, who was responsible for many of the experiments, tells me (L. Birnbaumer, personal communication) that they were pleased when the EC50 for glucagon's activation of adenyl cyclase appeared identical to its Kd for binding to membrane sites. But Lutz reminded his colleagues that the cyclase assay contained Mg^{2+} and ATP, while the binding assay did not. Repeating the binding as-

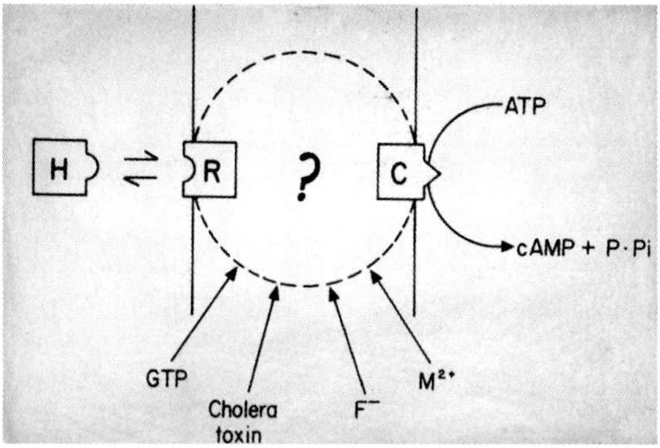

Fig. 1. The receptor and adenylyl cyclase in the early 1970s. The diagram is taken from a slide presented in seminars by the author in 1973–1975

says in the presence of Mg^{2+} and ATP produced a disconcerting result: the glucagon binding curve shifted to the right, with a higher Kd. Astutely, they tested other nucleotides: GTP shifted the binding curve more potently than ATP (Rodbell et al. 1971b). (The "pure" ATP they used turned out later to be contaminated by GTP.) A chemically pure synthetic ATP analog did not shift the binding curve, but did serve as an effective substrate for glucagon-stimulated cAMP synthesis, but only if GTP was added to the assay (Rodbell et al. 1971a).

These observations triggered fanciful speculations, but investigators were slow to realize that the evidence might point to a GTP-binding protein distinct from both receptor and adenylyl cyclase. Now we know that GTP reduced the receptor's affinity for glucagon by preventing the trimeric G-protein, Gs, from enhancing the GPCR's affinity for agonist: agonist affinity was reduced because GTP binding to Gs caused it to dissociate from the GPCR (De Lean et al. 1980; Ross and Gilman, 1980). At the time, however, many were not even convinced, despite accumulating evidence, that receptors and adenylyl cyclase were separate molecules (see Fig. 1). In 1975, Al Gilman's laboratory summarized their failed attempts to purify adenylyl cyclase in the title of a paper: "Frustration and adenylate cyclase" (Maguire et al. 1975).

Key insights into the mysterious relation between GTP and adenylyl cyclase came from Zvi Selinger's laboratory (Cassel et al. 1977;

Cassel and Selinger, 1976). He and his colleagues were intrigued by a report from the Rodbell laboratory (Londos et al. 1974) showing that a hydrolysis-resistant GTP analog, Gpp(NH)p, activated adenylyl cyclase on its own, and to an extent greater than GTP; moreover, Gpp(NH)p cooperated with hormones to further stimulate cAMP synthesis. If resistance to hydrolysis made GTP more effective, they reasoned that hormones might regulate GTPase activity. Soon the Selinger lab found that a GTPase activity in turkey erythrocyte membranes was stimulated by isoproterenol, and that this stimulation was blocked by propranolol. They proposed that cAMP synthesis depended on agonist-stimulated binding of GTP to a component of the adenylyl cyclase complex, that GTP hydrolysis terminated stimulation, and that continued cAMP synthesis required repeated agonist-stimulated cycles of GTP binding and hydrolysis. Their proposals were not greeted with enthusiasm. The *Journal of Biological Chemistry* rejected the first Selinger paper, which was deemed "prejudice not science", because "if anything, the hormone should inhibit GTP hydrolysis" (Z. Selinger, personal communication). In 1976, a respected senior investigator—perhaps a reviewer of the Selinger paper—admonished me to "be very cautious about accepting such a strange interpretation".

2 The Stimulatory Regulator of Adenylyl Cyclase

In 1972, I struck up a commute bus conversation with Gordon Tomkins, a UCSF faculty member. Gordon told me that somatic genetics—an entire field that was news to me—could furnish valuable clues to understanding hormone action. A postdoc in his laboratory had found that S49 mouse lymphoma cells die when exposed to a cAMP analog, and was beginning to isolate cAMP-resistant S49 variants (Daniel et al. 1973). cAMP resistance, Gordon suspected, resulted from mutation of a gene encoding a key protein in the cAMP response pathway. I jumped at the chance to join the project.

Soon I found myself working with Phil Coffino, an immensely talented postdoc in Gordon's lab. We isolated cAMP-resistant clones carrying mutations that inactivated protein kinase A (Bourne et al. 1975b; Coffino et al. 1975; Insel et al. 1975). Then we looked for an S49 clone lacking the β-adrenoceptor (β-AR). We imagined that such cells would die in the presence of cAMP analogs but resist killing by isoproterenol. To our surprise, clones that met these criteria also failed to die, or even to accumulate cAMP, in response to two additional stimulators of adenylyl cyclase, prostaglandin-E_1 and cholera toxin. We called these cells *cyc*$^-$, to indicate a deficiency of adenylyl cyclase (Bourne et al. 1975a), unaware that somatic genetics was hinting at existence of a protein we could not then imagine.

Later in 1974, Gordon received a postdoctoral application from a Cornell graduate student, Elliott Ross. Elliott's letter proposed to reconstitute hormone-sensitive adenylyl cyclase in *cyc*$^-$ membranes, using wild type S49 membranes as a source for purifying the component missing in *cyc*$^-$. Gordon promptly invited Elliott to join his lab, but it was not to be: a few months later, Gordon died after a brain operation, and Elliott joined Al Gilman's laboratory instead. We had sent *cyc*$^-$ cells to the Gilman laboratory as part of a separate collaboration, resulting in a paper (Insel et al. 1976) whose title revealed the meager state of our knowledge: "β-adrenergic receptors and adenylate cyclase: Products of separate genes?" (We got the right answer, all the while ignoring the fact that *cyc*$^-$ cells are not deficient in adenylyl cyclase.)

Much more important, with *cyc*$^-$ cells in hand Elliott could begin to tackle reconstitution of isoproterenol-stimulated adenylyl cyclase. It was not easy. Elliott and Al plowed through myriad detergent extractions and reconstitution strategies before they showed that *cyc*$^-$ can be persuaded—by addition of a membrane extract from wild type cells—to synthesize cAMP in response to isoproterenol (Ross and Gilman 1977a). Then came the critical observations: *cyc*$^-$ membranes do not lack adenylyl cyclase, and wild type extracts supplied to the reconstituted mixture an activity that was neither adenylyl cyclase nor the β-AR, both of which were already present in *cyc*$^-$; instead, the wild type extract supplied a new entity, whose thermal stability was increased by a GTP analog (Ross and Gilman 1977b).

By 1980 painstaking efforts in the Gilman laboratory had purified this entity, showing that the *cyc*$^-$ mutation inactivates a protein they named Gs, the stimulatory regulator of adenylyl cyclase (Ross and Gilman 1980). Discovery of the αβγ structure of Gs led rapidly to new insights, including the pathogenesis of three diseases. The ability of whooping cough (pertussis) toxin to inhibit GTP-dependent hormonal inhibition of adenylyl cyclase and catalyze covalent modification of a Gα protein distinct from αs (Murayama and Ui 1983) led to discovery and purification of a second putative trimeric G-protein, which we now call Gi (reviewed in Gilman 1987). (Now we know that this effect of pertussis toxin inhibits Gi activation, thereby causing the bronchial irritability of whooping cough.) Gαs, the target of the *cyc*$^-$ mutation, turned out to be the target of two diseases. Cholera is caused by a toxin that elevates cAMP in gut cells by covalently modifying αs, thereby turning off its GTPase activity and stabilizing it in its active form (Cassel and Pfeuffer 1978; Cassel and Selinger 1977; Johnson et al. 1978). Mutational inactivation of one αs allele causes the second disorder, pseudohypoparathyroidism, in which patients respond poorly to hormones that activate Gs-coupled receptors (Farfel et al. 1980).

3 Rhodopsin and Transducin

The extraordinary abundance of rhodopsin and transducin in retinal rod cells facilitated their initial discovery, and eventually made them the best-understood receptor-G-protein pair, at the levels of 3D structure, biochemical properties, and downstream signals. Rhodopsin was identified as a photosensitive pigment in the 1870s (reviewed in Hsia 1965), and in 1933 George Wald discovered retinal, rhodopsin's covalently bound ligand, and began to trace its light-induced chemical transformations (reviewed in Wald 1968). While the Gs and transducin stories evolved during the same time frame (Table 1), many aficionados of adenylyl cyclase and photoreception were barely aware of each other's findings until about 1980.

The transducin story began with three key findings: cGMP phosphodiesterase (PDE) was shown to be the light-activated effector (Bitensky et al. 1975); light increased the phosphodiesterase activity only in the presence of GTP and photoactivated rhodopsin (Yee and Liebman 1978); and light activated a GTPase activity in rod cell extracts (Wheeler and Bitensky 1977). Then Godchaux and Zimmerman (1979) purified from rod cell extracts a soluble guanine nucleotide binding protein that exhibited light-dependent stimulation of GTP-GDP exchange

Table 1 Trimeric G proteins and GPCRs: Key steps in discovery

	Adenylyl cyclase regulation	Phototransduction
Receptor identified	β-AR inferred from selectivity of agonists and antagonists (Ahlquist 1948)	1870s (reviewed in Hsia 1965)
Ligand identified	Epinephrine activates liver phosphorylase (Rall et al. 1957)	Retinal found by George Wald in 1933 (reviewed in Wald 1968)
Effector identified	Adenylyl cyclase (Sutherland and Rall 1960)	cGMP-PDE (Bitensky et al. 1975)
GTP	Regulates agonist binding and adenylyl cyclase activity (Rodbell et al. 1971a; Rodbell et al. 1971b)	Required for hν-stimulation of cGMP-PDE (Yee and Liebman 1978)
GTP hydrolysis	Stimulated by isoproterenol in turkey erythrocyte membranes (Cassel and Selinger 1976)	Activated by hν (Wheeler and Bitensky 1977)
Trimeric G protein purified	Gs α and β subunits (Northup et al. 1980)	Gt α and β (Godchaux and Zimmerman 1979); α, β, and γ (Fung et al. 1981)
Ligand binding	β-AR binds radioactive antagonists (Aurbach et al. 1974; Lefkowitz et al. 1974; Levitzki et al. 1974)	Retinal covalently attached to a lysine in rhodopsin (Collins, Morton, and Pitt 1950s; reviewed in Wald 1968)
Receptor phosphorylation	Desensitizes frog β-AR (Stadel et al. 1983)	hν-dependent (Kuhn and Dreyer 1972); role in adaptation (Kuhn et al. 1977)
Arrestin damps agonist signal	Retinal arrestin inhibits β-AR signaling (Benovic et al. 1987)	Retinal arrestin quenches hν-dependent cGMP-PDE activation (Wilden et al. 1986)
Receptor kinase	βARK (Benovic et al. 1986)	Rhodopsin kinase (Kuhn 1978; Shichi and Somers 1978)

Table 1 (continued)

	Adenylyl cyclase regulation	Phototransduction
Receptor purified	β-AR, by affinity chromatography (Benovic et al. 1984; Caron et al. 1979; Shorr et al. 1981)	1960s, many labs (reviewed in Hsia 1965)
Gα cDNA	αs (Harris et al. 1985)	αt (Lochrie et al. 1985; Medynski et al. 1985; Tanabe et al. 1985)
GPCR primary structure	β_2-AR cDNA (Dixon et al. 1986; Yarden et al. 1986)	Rhodopsin amino acid sequence (Ovchinnikov et al. 1982); cDNA (Nathans and Hogness 1983)

in the presence of membranes. They identified two polypeptide components of the soluble protein, which we now know as the α and β subunits of transducin, but did not mention adenylyl cyclase or GTP's role in its hormonal activation.

By 1981, reports from Lubert Stryer's laboratory made it impossible to ignore striking parallels between retinal phototransduction and hormone-stimulated cAMP synthesis. Lubert and his colleagues showed that the photon signal is enormously amplified: a single photon, activating a single rhodopsin, triggers binding of a hydrolysis-resistant GTP analog to 500 GTP-binding sites in rod cell extracts (Fung and Stryer, 1980). They then purified the GTP-binding protein, identified its α, β, and γ polypeptides, named it transducin (hereafter, Gt), and used it and rhodopsin to reconstitute light-stimulated binding and hydrolysis of GTP (Fung et al. 1981).

At this point, the two previously unrelated fields of investigation began to coalesce, each providing knowledge and insights to the other. Gs and Gt would quickly give rise to a larger family of trimeric G-proteins as well as a growing retinue of effectors and auxiliary regulators (for examples, see Table 1). Why then did I (and, I suspect, many of my colleagues) find family resemblances between Gs and Gt so surprising in 1980? One reason may be that laboratories focusing on different problems communicated less often with one another in 1980 than they do in the 21st century. More likely, we were simply not ready to imagine close parallels between disparate biological functions: why, after all, should cells in the liver and retina use nearly identical machinery to detect glucagon vs photons? Now such a revelation would not come as a surprise, because we have learned that evolution makes each new signaling machine by modifying and cobbling together parts of machines already in use somewhere else. For many of us, Gs and Gt furnished the first inkling of this principle.

4 Confusion, Error, Truth: Discovering the β-AR

By the early 1970s, investigators were beginning to transform putative hormone receptors into biochemical entities by binding radioactive agonist peptides to receptors in tissue extracts. In 1948, Ahlquist had postulated the existence of two classes of catecholamine receptors, which he called α and β (Ahlquist 1948). Bob Lefkowitz and Gerry Aurbach, among others, saw a straightforward route to identifying these receptors: assess binding of ^{3}H-labelled catecholamines to particulate extracts of tissues with catecholamine-sensitive adenylyl cyclase activity. Unfortunately, a good idea may stir up confusion rather than shed light. The β-AR story unfolded much as predicted by the pioneer of scientific induction 400 years ago: "... truth will sooner come out from error than from confusion," wrote Francis Bacon in his *Novum Organum* (1620).

First came an era of confusion: in the early 1970s the Aurbach and Lefkowitz laboratories found plenty of ^{3}H-norepinephrine binding sites, with binding that was usually reversible and competed by nonradioactive catecholamine agonists (Bilezikian and Aurbach 1973a; Lefkowitz and Haber 1971); some reports even claimed receptor solubilization, affinity chromatography, and partial purification (Bilezikian and Aurbach 1973b; Lefkowitz 1973; Lefkowitz et al. 1972). These investigators also found disturbing mismatches between patterns of agonist binding and response: agents without agonist or antagonist activity, such as inactive optimal isomers of norepinephrine or dihydroxymandelic acid, efficiently competed against ^{3}H-norepinephrine for binding, while β-AR antagonists such as propranolol competed poorly, even at concentrations orders of magnitude greater than propranolol's IC50 (summarized in Lefkowitz 1974).

These discrepancies led to fanciful interpretations: perhaps the antagonist first associates with a necessary-but-not-sufficient "partial" binding site but does not activate the receptor unless it also interacts with one or more additional sites; the first site would be detected by binding of ^{3}H-norepinephrine, the second only by receptor activation (Bilezikian and Aurbach 1973a; Lefkowitz 1974). One review even suggested that perhaps there were "certain inherent limitations in relying solely on the criteria of specificity and affinity of binding for identification of receptors" (Lefkowitz 1974). The same review admitted, however, that "the data available ... are not ... sufficient to prove or disprove the hypothesis that these [binding] sites represent the β-adrenergic receptor binding sites." It was beginning to dawn on investigators that their confusion might reflect what Francis Bacon referred to as "error."

This recognition allowed truth to emerge from error. Maguire and co-workers (1974) showed that ascorbic acid and sodium metabisulfite prevented ^{3}H-norepinephrine binding, suggesting that the binding represented covalent attachment of oxidized radioactive products to macromolecules other than receptors. In the same year, three laboratories (Aurbach et al. 1974; Lefkowitz et al. 1974; Levitzki et al. 1974) reported that non-catechol β-AR antagonists bind to sites with specificities for competition by optical isomers, agonists, and other antagonists that match those expected for the real β-AR.

Reliable binding assays for β-ARs allowed their biochemical characterization and eventual purification. Because biochemistry can be hard, the new "truth" did not make further advances easy. Undaunted, Caron, Lefkowitz, and their colleagues eventually purified detergent-solubilized β-AR protein by affinity chromatography (Benovic et al. 1984; Caron et al. 1979; Shorr et al. 1981). Availability of pure receptor protein soon made it possible to reconstitute pure β-AR with Gs and adenylyl cyclase (Cerione et al. 1984; May et al. 1985) and to identify β-AR kinase (Benovic et al. 1986). Most important, the β-AR story was developing in the period when recombinant DNA technology was beginning to hit its stride. Pure receptors made it possible to probe genomic DNA libraries with nucleotide probes based on receptor peptides. Amino acid sequences of β_2-ARs from hamster and from turkey erythrocytes led to cloning receptor cDNAs from these animals and predictions of the receptors' very similar amino acid sequences (Dixon et al. 1986; Yarden et al. 1986).

Some of us still remember the enormous excitement generated by the obvious similarities between primary structures of rhodopsin (Nathans and Hogness 1983) and the β_2-AR (Dixon et al. 1986; Yarden et al. 1986). The seven homologous hydrophobic α helices heralded the birth of a GPCR superfamily. Our delighted surprise paralleled the surprise generated by the discoveries of Gs and Gt. Again we had failed to anticipate evolution's propensity to adapt a successful piece of machinery to new uses. Delight and surprise were even greater this time, because cDNA sequences of αs and αt had just been reported (see Table 1). For us, the Gα and GPCR primary structures were harbingers of a torrent of new discoveries, driven by the power of molecular biology.

5 DNA: Revolution and Revelation

Rather than attempt a comprehensive account of the dazzling post-DNA history of G-proteins and GPCRs, I shall end this essay with glancing sketches of a few examples from this history, and point out how they have altered our ways of posing and solving questions. Pre-DNA discoveries contained the essential seeds of a series of new general concepts (indicated below in italics). Without the DNA revolution, however, none of these would have reached its present level of explanatory power. Now each of these ideas is an essential item of an investigator's intellectual furniture, necessary for designing and interpreting almost every experiment.

One such general concept is that of the *regulatory protein module*. The versatile R-G-E triad, comprising a GPCR, a trimeric G-protein, and an effector, is one of the best-studied regulatory modules in biology. The striking biochemical parallels between regulation of cAMP synthesis and phototransduction, in combination with similar primary structures of αs *vs* αt and of β-ARs vs rhodopsin, made R-G-E one of the very first of these modules. This module, we now know, is responsible for the mating dance of yeast and for detecting sensory cues and intercellular signals in flies, worms, mice, and humans. To see how far we have come, contrast the puzzle of hormone-sensitive adenylyl cyclase in the 1970s (Fig. 1) with the crystal-clear atomic structures of triad members solved two decades later: rhodopsin's transmembrane helices (Fig. 2a) and a complex of αs with adenylyl cyclase (Fig. 2b).

Like MAP kinase cascades, cytokine receptor signaling via JAK/STAT complexes, and many other modules, the R-G-E module is a set of evolutionarily conserved proteins that uses a common mechanism to transduce signals between different sets of inputs and outputs. From our standpoint in the 21st century, it may seem extraordinary that the concept of regulatory modules required a major shift in our way of looking at the world. In essence, we rediscovered evolution. Before DNA sequences came on the scene, scientists tended to imagine that *their* question and the molecule they hoped would answer it were essentially unique. In contrast, the R-G-E module showed us that duplication and divergence of GPCR and GTPase genes, combined by selection of useful gene products, had produced a module with interchangeable subunits

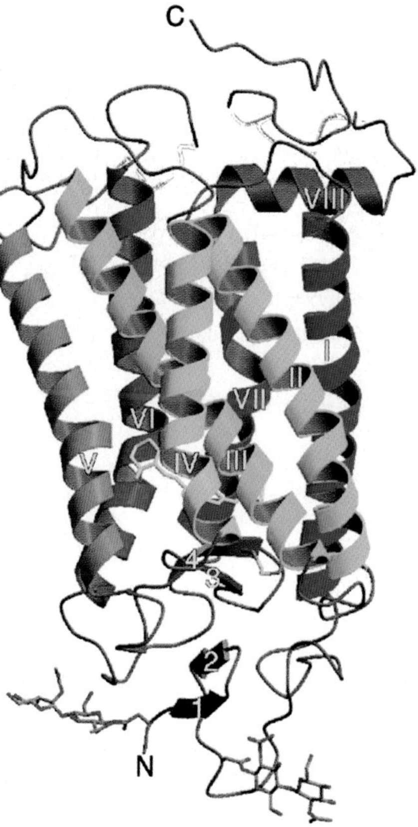

a

Fig. 2a,b. Atomic structures of a GPCR and a Gα-effector complex. **a** Rhodopsin, showing the seven transmembrane helices (*colored and numbered with Roman numerals*), loops connecting them (extracellular at *bottom*, cytoplasmic at *top*), and retinal (*yellow*). (Reprinted with permission from Fig. 2A of Palczewski et al. 2000, Science 289:739–745; copyright 2000 AAAS). **b** The α subunit of Gs (*left*) interacting with the catalytic domains of adenylyl cyclase (*right*). (Reprinted with permission from Fig. 4 of Tesmer et al. 1997; Science 278:1907–1916; copyright 1997 AAAS)

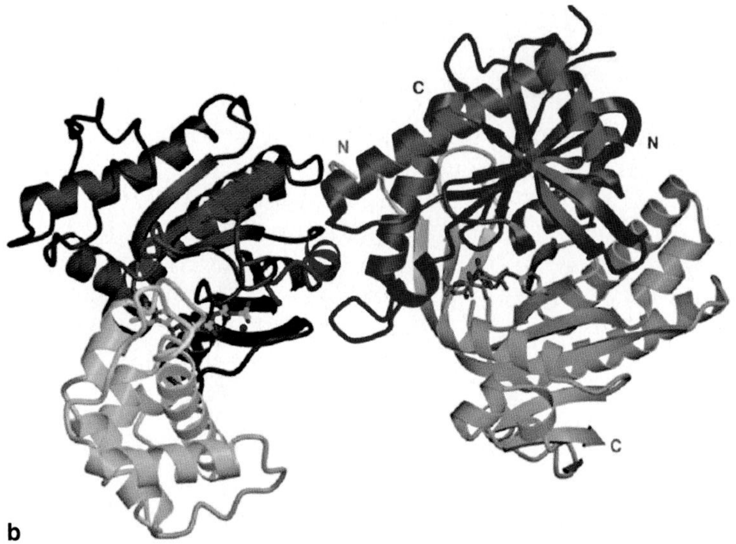

Fig. 2a,b. (continued)

that can selectively link large numbers of distinct inputs to different outputs.

The DNA revolution also created the closely related idea of *protein families*. Growing families and subfamilies of GPCRs and G-proteins brought to light hundreds of targets for intensive research in hormone action, vision, olfaction, neurobiology, immune responses, and embryonic development. A bevy of intriguing orphan GPCRs stands ready to join their ranks. Conserved regions of primary structure in other protein families revealed families of auxiliary proteins (e.g., RGS and Goloco) that interact with the R-G-E module. Gα subunits share sequence and three-dimensional architecture with a huge superfamily of GTPase switches, which also includes bacterial elongation factors, Ras, a host of other small GTPases, and many others. Evolution found that a good switch is worth conserving.

By linking R-G-E modules to other regulatory proteins (PDEs, kinases, phosphatases, ion channels, and more), AKAPs and other scaffolds create *higher-order protein complexes*, which in turn harness specific stimuli to an enormous variety of responses. In 1980, allostery and covalent modification were recognized as the principal modes of signal transduction. To them we now add a third, just as essential: *regulated proximity of proteins and signaling modules.*

From pheromone receptors in yeast to rhodopsin and chemokine receptors in vertebrates, GPCR activation triggers densely complex regulatory circuits, replete with positive and negative feedback loops. We can now begin to trace and manipulate such *cellular signaling networks* in space and time, using recombinant fluorescent probes, mRNA arrays, RNAi, the polymerase chain reaction, genomic sequences of many animals, and a host of other new tools. Without these it would be impossible to measure—or even to conceive—physiologically crucial temporal or spatial changes in the interactions of GPCRs, arrestins, or effector substrates and products (e.g., PIP2 or PIP3) with one another.

Discoveries at the atomic level include the conserved architecture and molecular mechanism of the conformational switch common to small GTPases and Gα subunits; interactions of G-protein subunits with effectors and other regulators; and how one GPCR ligand, 11-*cis*-retinal, nestles within the seven-helix bundle of its receptor, rhodopsin (Fig. 2a). All but the last of these discoveries depended on modifying and expressing recombinant genes. As a result, *regulation at the level of conformational change* (aka allostery) is no longer confined to a few molecules such as hemoglobin and conceptual models of other molecules; instead, documented conformational change regularly generates testable hypotheses and experiments.

Although I have focused on G-protein and GPCR research, every discovery I mention has myriad counterparts in virtually every field of present-day biomedical research. Consequently, molecular biology's rapidly expanding toolbox and the new ideas it generates have dramatically altered our laboratories, how we interact with each other, and our goals and expectations. Laboratories are larger and depend on much more powerful technology. Even the disposable plastic tips of today's ubiquitous pipette-man would have amazed experimenters who depended on individually calibrated glass lambda pipettes, operated by sucking on a rubber tube and meticulously washed with acid after each use. For the average investigator, scientific communication is faster, and critically important research papers and seminars more frequent. In the early 1970s, one meeting per year was often more than enough. Now we are much more frequently thrilled (or disconcerted) by a new finding directly pertinent to the question we are asking, and suddenly find ourselves learning a new technology or immersed in a whole new field.

For researchers today, these exciting changes have produced two especially wide-ranging consequences. First, we justifiably expect our research to produce more rapid and far-ranging discoveries. We complain mightily, of course, about funding, bureaucracy, competition, failed experiments, and threatening social or political developments, just as we did in the 1970s. More significantly, we now feel reasonably sure that tomorrow we will understand more than we do today.

The second consequence is closely related to these changed expectations and even more crucial: investigators now expect their discoveries to prove relevant and even genuinely useful in the world outside the laboratory. As compared to the days when G-proteins and GPCRs were born, individual scientists and ideas travel much more rapidly and efficiently between basic and clinical science, and between academia and the pharmaceutical industry.

Although expectations do not tell us what the future will bring, I find it encouraging to look back to the birth of our field. The questions scientists posed in 1970 led eventually to today's discoveries, and more questions, none of which any of us could have imagined in our wildest dreams.

Acknowledgements. I thank Thomas Sakmar, Lutz Birnbaumer, Paul Insel, Zvi Selinger, and Elliott Ross for commenting on parts of the manuscript. Mistakes, however, are clearly my own.

References

Ahlquist R (1948) A study of the adrenergic receptors. Am J Physiol 153:586–600

Aurbach GD, Fedak SA, Woodard CJ, Palmer JS, Hauser D, Troxler F (1974) Beta-adrenergic receptor: stereospecific interaction of iodinated beta-blocking agent with high affinity site. Science 186:1223–1224

Benovic JL, Kuhn H, Weyand I, Codina J, Caron MG, Lefkowitz RJ (1987) Functional desensitization of the isolated beta-adrenergic receptor by the beta-adrenergic receptor kinase: potential role of an analog of the retinal protein arrestin (48-kDa protein). Proc Natl Acad Sci USA 84:8879–8882

Benovic JL, Shorr RG, Caron MG, Lefkowitz RJ (1984) The mammalian beta 2-adrenergic receptor: purification and characterization. Biochemistry 23: 4510–4518

Benovic JL, Strasser RH, Caron MG, Lefkowitz RJ (1986) Beta-adrenergic receptor kinase: identification of a novel protein kinase that phosphorylates the agonist-occupied form of the receptor. Proc Natl Acad Sci USA 83:2797–2801

Bilezikian JP, Aurbach GD (1973a) A beta-adrenergic receptor of the turkey erythrocyte. I. Binding of catecholamine and relationship to adenylate cyclase activity. J Biol Chem 248:5577–5583

Bilezikian JP, Aurbach GD (1973b) A beta-adrenergic receptor of the turkey erythrocyte. II. Characterization and solubilization of the receptor. J Biol Chem 248:5584–5589

Bitensky MW, Miki N, Keirns JJ, Keirns M, Baraban JM, Freeman J, Wheeler MA, Lacy J, Marcus FR (1975) Activation of photoreceptor disk membrane phosphodiesterase by light and ATP. Adv Cyclic Nucleotide Res 5:213–240

Bourne HR, Coffino P, Tomkins GM (1975a) Selection of a variant lymphoma cell deficient in adenylate cyclase. Science 187:750–752

Bourne HR, Coffino P, Tomkins GM (1975b) Somatic genetic analysis of cyclic AMP action: characterization of unresponsive mutants. J Cell Physiol 85:611–620

Caron MG, Srinivasan Y, Pitha J, Kociolek K, Lefkowitz RJ (1979) Affinity chromatography of the beta-adrenergic receptor. J Biol Chem 254:2923–2927

Cassel D, Levkovitz H, Selinger Z (1977) The regulatory GTPase cycle of turkey erythrocyte adenylate cyclase. J Cyclic Nucleotide Res 3:393–406

Cassel D, Pfeuffer T (1978) Mechanism of cholera toxin action: covalent modification of the guanyl nucleotide-binding protein of the adenylate cyclase system. Proc Natl Acad Sci USA 75:2669–2673

Cassel D, Selinger Z (1976) Catecholamine-stimulated GTPase activity in turkey erythrocyte membranes. Biochim Biophys Acta 452:538–551

Cassel D, Selinger Z (1977) Mechanism of adenylate cyclase activation by cholera toxin: inhibition of GTP hydrolysis at the regulatory site. Proc Natl Acad Sci USA 74:3307–3311

Cerione RA, Sibley DR, Codina J, Benovic JL, Winslow J, Neer EJ, Birnbaumer L, Caron MG, Lefkowitz RJ (1984) Reconstitution of a hormone-sensitive adenylate cyclase system. The pure beta-adrenergic receptor and guanine nucleotide regulatory protein confer hormone responsiveness on the resolved catalytic unit. J Biol Chem 259:9979–9982

Coffino P, Bourne HR, Tomkins GM (1975) Somatic genetic analysis of cyclic AMP action: selection of unresponsive mutants. J Cell Physiol 85:603–610

Daniel V, Litwack G, Tomkins GM (1973) Induction of cytolysis of cultured lymphoma cells by adenosine 3′:5′-cyclic monophosphate and the isolation of resistant variants. Proc Natl Acad Sci USA 70:76–79

De Lean A, Stadel JM, Lefkowitz RJ (1980) A ternary complex model explains the agonist-specific binding properties of the adenylate cyclase-coupled beta-adrenergic receptor. J Biol Chem 255:7108–7117

Dixon RA, Kobilka BK, Strader DJ, Benovic JL, Dohlman HG, Frielle T, Bolanowski MA, Bennett CD, Rands E, Diehl RE et al. (1986) Cloning of the gene and cDNA for mammalian beta-adrenergic receptor and homology with rhodopsin. Nature 321:75–79

Farfel Z, Brickman AS, Kaslow HR, Brothers VM, Bourne HR (1980) Defect of receptor-cyclase coupling protein in pseudohypoparathyroidism. New Engl J Med 303:237–242

Fung B, Stryer L (1980) Photolyzed rhodopsin catalyzes the exchange of GTP for bound GDP in retinal rod outer segments. Proc Natl Acad Sci USA 77:2500–2504

Fung BK, Hurley JB, Stryer L (1981) Flow of information in the light-triggered cyclic nucleotide cascade of vision. Proc Natl Acad Sci USA 78:152–156

Gilman AG (1987) G proteins: transducers of receptor-generated signals. Annu Rev Biochem 56:615–649

Godchaux W, Zimmerman WF (1979) Membrane-dependent guanine nucleotide binding and GTPase activities of soluble protein from bovine rod cell outer segments. J Biol Chem 254:7874–7884

Harris BA, Robishaw JD, Mumby SM, Gilman AG (1985) Molecular cloning of complementary DNA for the alpha subunit of the G protein that stimulates adenylate cyclase. Science 229:1274–1277

Hsia Y (1965) Photochemistry of vision. In: Graham CH (ed) Vision and visual perception. Wiley, New York

Insel PA, Bourne HR, Coffino P, Tomkins GM (1975) Cyclic AMP-dependent protein kinase: pivotal role in regulation of enzyme induction and growth. Science 190:896–898

Insel PA, Maguire ME, Gilman AG, Bourne HR, Coffino P, Melmon KL (1976) Beta adrenergic receptors and adenylate cyclase: products of separate genes? Mol Pharmacol 12:1062–1069

Johnson GL, Kaslow HR, Bourne HR (1978) Genetic evidence that cholera toxin substrates are regulatory components of adenylate cyclase. J Biol Chem 253:7120–7123

Kuhn H (1978) Light-regulated binding of rhodopsin kinase and other proteins to cattle photoreceptor membranes. Biochemistry 17:4389–4395

Kuhn H, Dreyer WJ (1972) Light dependent phosphorylation of rhodopsin by ATP. FEBS Lett 20:1–6

Kuhn H, McDowell JH, Leser KH, Bader S (1977) Phosphorylation of rhodopsin as a possible mechanism of adaptation. Biophys Struct Mech 3:175–180

Lefkowitz RJ (1973) Isolated hormone receptors: physiologic and clinical implications. N Engl J Med 288:1061–1066

Lefkowitz RJ (1974) Commentary. Molecular pharmacology of beta-adrenergic receptors: a status report. Biochem Pharmacol 23:2069–2076

Lefkowitz RJ, Haber E (1971) A fraction of the ventricular myocardium that has the specificity of the cardiac beta-adrenergic receptor. Proc Natl Acad Sci USA 68:1773–1777

Lefkowitz RJ, Haber E, O'Hara D (1972) Identification of the cardiac beta-adrenergic receptor protein: solubilization and purification by affinity chromatography. Proc Natl Acad Sci USA 69:2828–2832

Lefkowitz RJ, Mukherjee C, Coverstone M, Caron MG (1974) Stereospecific (3H)(minus)-alprenolol binding sites, beta-adrenergic receptors and adenylate cyclase. Biochem Biophys Res Commun 60:703–709

Levitzki A, Atlas D, Steer ML (1974) The binding characteristics and number of beta-adrenergic receptors on the turkey erythrocyte. Proc Natl Acad Sci USA 71:2773–2776

Lochrie MA, Hurley JB, Simon MI (1985) Sequence of the alpha subunit of photoreceptor G protein: homologies between transducin, ras, and elongation factors. Science 228:96–99

Londos C, Salomon Y, Lin MC, Harwood JP, Schramm M, Wolff J, Rodbell M (1974) 5′-Guanylylimidodiphosphate, a potent activator of adenylate cyclase systems in eukaryotic cells. Proc Natl Acad Sci USA 71:3087–3090

Maguire ME, Goldmann PH, Gilman AG (1974) The reaction of [3H]norepinephrine with particulate fractions of cells responsive to catecholamines. Mol Pharmacol 10:563–581

Maguire ME, Sturgill TW, Gilman AG (1975) Frustration and adenylate cyclase. Metabolism 24:287–299

May DC, Ross EM, Gilman AG, Smigel MD (1985) Reconstitution of catecholamine-stimulated adenylate cyclase activity using three purified proteins. J Biol Chem 260:15829–15833

Medynski DC, Sullivan K, Smith D, Van Dop C, Chang FH, Fung BK, Seeburg PH, Bourne HR (1985) Amino acid sequence of the alpha subunit of transducin deduced from the cDNA sequence. Proc Natl Acad Sci USA 82:4311–4315

Murayama T, Ui M (1983) Loss of the inhibitory function of the guanine nucleotide regulatory component of adenylate cyclase due to its ADP ribosylation by islet-activating protein, pertussis toxin, in adipocyte membranes. J Biol Chem 258:3319–3326

Nathans J, Hogness DS (1983) Isolation, sequence analysis, and intron-exon arrangement of the gene encoding bovine rhodopsin. Cell 34:807–814

Northup JK, Sternweis PC, Smigel MD, Schleifer LS, Ross EM, Gilman AG (1980) Purification of the regulatory component of adenylate cyclase. Proc Natl Acad Sci USA 77:6516–6520

Ovchinnikov YA, Abdulaev NG, Feigina MY, Artamonov ID, Zolotarev AS, Miroshnikov AL, Martynov VL, Kostina MB, Kudelin AB, Bogachuk AS (1982) The complete amino acid sequence of visual rhodopsin. Bioorg Khim 8:1424–1427

Palczewski K, Kumasaka T, Hori T, Behnke CA, Motoshima H, Fox BA, Le Trong I, Teller DC, Okada T, Stenkamp RE, Yamamoto M, Miyano M (2000) Crystal structure of rhodopsin: A G protein-coupled receptor. Science 289:739–745

Rall TW, Sutherland EW, Berthet J (1957) The relationship of epinephrine and glucagon to liver phosphorylase. IV. Effect of epinephrine and glucagon on the reactivation of phosphorylase in liver homogenates. J Biol Chem 224:463–475

Rodbell M, Birnbaumer L, Pohl SL, Krans HMJ (1971a) The glucagon-sensitive adenyl cyclase system in plasma membranes of rat liver. V. An obligatory role of guanyl nucleotides in glucagon action. J Biol Chem 246:1877–1882

Rodbell M, Krans HMJ, Pohl SL, Birnbaumer L (1971b) The glucagon-sensitive adenyl cyclase system in plasma membranes of rat liver. IV. Effects of guanyl nucleotides on binding of ^{125}I-glucagon. J Biol Chem 246:1872–1876

Ross EM, Gilman AG (1977a) Reconstitution of catecholamine-sensitive adenylate cyclase activity: interactions of solubilized components with receptor-replete membranes. Proc Natl Acad Sci USA 74:3715–3719

Ross EM, Gilman AG (1977b) Resolution of some components of adenylate cyclase necessary for catalytic activity. J Biol Chem 252:6966–6969

Ross EM, Gilman AG (1980) Biochemical properties of hormone-sensitive adenylate cyclase. Annu Rev Biochem 49:533–564

Shichi H, Somers RL (1978) Light-dependent phosphorylation of rhodopsin. Purification and properties of rhodopsin kinase. J Biol Chem 253:7040–7046

Shorr RG, Lefkowitz RJ, Caron MG (1981) Purification of the beta-adrenergic receptor. Identification of the hormone binding subunit. J Biol Chem 256:5820–5826

Stadel JM, Nambi P, Shorr RG, Sawyer DF, Caron MG, Lefkowitz RJ (1983) Catecholamine-induced desensitization of turkey erythrocyte adenylate cyclase is associated with phosphorylation of the beta-adrenergic receptor. Proc Natl Acad Sci USA 80:3173–3177

Sutherland EW, Rall TW (1960) Formation of adenosine-3,5-phosphate (cyclic adenylate) and its relation to the action of several neurohormones or hormones. Acta Endocrinol (Copenh) 34(Suppl 50):171–174

Tanabe T, Nukada T, Nishikawa Y, Sugimoto K, Suzuki H, Takahashi H, Noda M, Haga T, Ichiyama A, Kangawa K et al. (1985) Primary structure of the alpha-subunit of transducin and its relationship to ras proteins. Nature 315:242–245

Tesmer JJ, Sunahara RK, Gilman AG, Sprang SR (1997) Crystal structure of the catalytic domains of adenylyl cyclase in a complex with Gsalpha. GTP-gammaS. Science 278:1907–1916

Wald G (1968) The molecular basis of visual excitation. Nature 219:800–807

Wheeler GL, Bitensky MW (1977) A light-activated GTPase in vertebrate photoreceptors: regulation of light-activated cyclic GMP phosphodiesterase. Proc Natl Acad Sci USA 74:4238–4242

Wilden U, Hall SW, Kuhn H (1986) Phosphodiesterase activation by photoexcited rhodopsin is quenched when rhodopsin is phosphorylated and binds the intrinsic 48-kDa protein of rod outer segments. Proc Natl Acad Sci USA 83:1174–1178

Yarden Y, Rodriguez H, Wong SK, Brandt DR, May DC, Burnier J, Harkins RN, Chen EY, Ramachandran J, Ullrich A et al. (1986) The avian beta-adrenergic receptor: primary structure and membrane topology. Proc Natl Acad Sci USA 83:6795–6799

Yee R, Liebman PA (1978) Light-activated phosphodiesterase of the rod outer segment. Kinetics and parameters of activation and deactivation. J Biol Chem 253:8902–8909

Ernst Schering Foundation Symposium Proceedings, Vol. 2, pp. 23–47
DOI 10.1007/2789_2006_002

Published Online: 16 May 2007

Modeling GPCRs

A.C.M. Paiva†, L. Oliveira, F. Horn†, R.P. Bywater, G. Vriend(✉)

CMBI NCMLS, UMC, Geert Grooteplein 28, 6525 GA Nijmegen, The Netherlands
email: *Vriend@CMBI.ru.nl*

In memoriam Antonio Paiva

While we were working on this article, our good friend, colleague, and mentor, Antonio Paiva, died after losing the fight to cancer. We know Antonio as a stimulating force in the GPCR field. He has been one of the founding fathers of the informal GPCR club that met regularly at the EMBL in the early 1990s. The GPCRDB sprouted from these meetings. The thousands of scientists that use the GPCRDB every day owe Antonio thanks. His co-authors will miss him, and send words of condolence to his family. May they find consolation in the fact that the cancer could not stop him from finishing this article. We will miss him.

In memoriam Florence Horn

This paper is a dedication to the memory of Florence Horn who died shortly after we finished this article. Flo did so much to make genomics data come alive in a way that was meaningful to her and to thousands of researchers in bioinformatics, molecular biology, structural biology, and medicinal chemistry.

Along with all these researchers, the coauthors of this paper wish to put on record their thanks to Flo and to remember her as a colleague who bestowed her good humor and joie de vivre on all those who worked with her.

Abstract. Many GPCR models have been built over the years for many different purposes, of which drug-design undoubtedly has been the most frequent one. The release of the structure of bovine rhodopsin in August 2000 enabled us to analyze models built before that period to learn things for the models we build today. We conclude that the GPCR modeling field is riddled with "common knowledge". Several characteristics of the bovine rhodopsin structure came as a big surprise, and had obviously not been predicted, which led to large errors in the models. Some of these surprises, however, could have been predicted if the modelers had more rigidly stuck to the rule that holds for all models, namely that a model should explain all experimental facts, and not just those facts that agree with the modeler's preconceptions.

1 Introduction

GPCRs are essential components in biological signaling processes in higher animals and accordingly, for humans, constitute the most important set of targets for the pharmaceutical industry, as is indicated by the fact that 52% of all medicines available today act on them (Watson and Arkinstall 1994). Approximately 16,700 GPCR sequences are publicly available today (Bairoch et al. 2005), including 1,795 human proteins. The GPCRDB (Horn et al. 1998, 2003) is a worldwide repository for GPCR-related data. In addition to sequence data and multiple sequence alignments, the GPCRDB (www.gpcr.org/7tm/) gives access to approximately 8,000 mutations (Beukers et al. 1999; Horn et al. 2004). Binding constants are available for approximately 30,000 ligand-receptor combinations obtained from two different sources. Massive data is also available regarding chromosomal location, cDNA sequences, secondary structure, 3D models, and correlated mutation analyses. Query and navigation tools are also provided and allow users to retrieve local and remote information such as associated disease states, localizations, post-translational modifications, etc. Snake-like diagrams (Campagne et al. 2003) are used to offer a two-dimensional view of the receptors but also to combine sequence, structure, and mutation data. In the database, the data organization is based on the pharmacological classification of GPCRs. In addition to the five main classes (A–E), other putative GPCR families are also described, these are the frizzled/smoothened family, ocular albinism proteins, insect odorant receptors, plant Mlo receptors, nematode chemoceptors, vomeronasal receptors, taste receptors T2R, as well as numerous unclassified receptors. Bacteriorhodopsins are present for historical reasons. It is worth noting that most of the GPCRs present in the GPCRDB have not (yet) been proven to couple to G-proteins and we should rather talk about heptahelical receptors—and maybe rename the database 7TMDB.

There have been many dramatic developments in the use of modern "omics" technologies in drug design from genomics to metabonomics. Nevertheless, the chemical structure/function space is both disjoint and replete with highly redundant structures, which makes navigation difficult. Still today an element of luck is necessary, and this is reflected in the fact that the rate of discovery of new medicines has declined.

Structure-based design has occasionally been successful, but it is precisely in the GPCR area where structure-based design has not worked satisfactorily. The paucity of accurate structural data for GPCR templates and the desire to remedy this situation has spawned an entire generation of modelers intent on calculating/predicting GPCR structures. Before August 4, 2000, bacteriorhodopsin (Henderson and Schertler 1990; Pebay-Peyroula et al. 1997; Luecke et al. 1998; Takeda et al. 1998) was often used as a modeling template, but on that date the three-dimensional coordinates (Palczewski et al. 2000) of bovine rhodopsin became available, providing a much better template for GPCR modeling than bacteriorhodopsin, which is not even a GPCR. Moreover, bovine rhodopsin is not the perfect template, as we will explain in this chapter. Models produced Before the Crystal structure became available are called BC models, and those produced After these Data became available, AD models.

1.1 BC Modeling

Most BC models were based on low-resolution electron cryomicroscopic models of bacteriorhodopsin (Henderson and Schertler 1990) while the precision (but not the accuracy) was improved when X-ray crystal structures of bacteriorhodopsin became available (Pebay-Peyroula et al. 1997; Luecke et al. 1998; Palczewski et al. 2000). The situation improved with the availability of the Cα coordinates produced by J. Baldwin (Baldwin 1993) from an electron diffraction map (Unger and Schertler 1995; Schertler et al. 1993; Unger et al. 1997; Schertler and Hargrave 1995) produced by the Schertler group. A few models (Filizola et al. 1998; Prusis et al. 1997; Bramblett et al. 1995) were based on first principles, sometimes guided by low-resolution data measured from published slices of the electron density maps for the bovine or frog rhodopsin.

The BC modeling community developed a series of dogmas that are summarized in Box 1. Many of these are unfortunately still applied to this day by some modelers.

Given these dogmas, it can easily be understood why most modeling recipes followed the steps listed in Box 2.

Box 1 GPCR modeling dogmas and misconceptions from the BC era

Loops stick out into the solvent.
Isolated loops have the same structure as in a GPCR.
Polar residues point inward.
Helices stop at the membrane surface.
All helices are about equally long.
Helices must be perfect.
Helices are organized in a semicircular fashion.
Molecular dynamics software improves models.
There must be space in the apo-form for a ligand.
Activation does not require motion.
Important residues bind ligand or G-protein.
Important residues point inward.
The bacteriorhodopsin structure is a solution to the problem of how to pack seven helices in the membrane. It is therefore the only solution.
Bacteriorhodopsin is a GPCR without G-protein.
The lysines in helix VII should line up.
Proteins are simple.
Models are correct.

Box 2 Typical steps in a generic BC modeling project

First	Determine which template to use, or design your own helix-packing model.
Second	Use threading or moment calculations to determine the mapping of the GPCR sequence onto the selected template. Moment calculations can be based on hydrophobic moments (Donnelly et al. 1993), conservation moments (Pardo et al. 1992), etc., or a combination of these (Herzyk and Hubbard 1998). Threading can be based on general rules, helix bundle rules (Herzyk and Hubbard 1998; Pogozheva et al. 1997), or even bacteriorhodopsin-specific rules (Cronet et al. 1993).
Third	Find experimental data that agree with the model and add them to convince yourself or the referees that this is the only correct model.

We found very many publications that discussed poor BC models, showing that bluff will fly with referees and editors if the topic is important enough. Sadly, even the poorest models seemed to agree with all data (selected by the authors) and seemed to be perfect for designing drugs (according to the authors). As a courtesy, we will not list those articles here.

1.2 The Bovine Rhodopsin Structure

The high-resolution structure of rhodopsin (Palczewski et al. 2000; Schertler 2005) reveals a seven-helix bundle with a central cavity surrounded by helices I–III and V–VII (see Fig. 1). The helices are in blue-purple. The β-hairpin in the N-terminal domain and the β-hairpin between helices IV and V (commonly known as the second extracellular loop) are in orange. The retinal is in yellow. Irregular parts are in blue-green. The topmost helix is helix IV.

Helix IV is not part of the cavity wall in this structure and makes contacts only with helix III. However, helix IV has been suggested to make contacts with some agonists which, if correct, is one of the many pieces of evidence that the active structure differs from the inactive one represented here. The conserved tryptophan at position 420 (we use GPCRDB residue numbering throughout this article) in helix IV is far away from any other residue known to have a functional role. This tryptophan might therefore play a role in receptor dimerization. Dimeriza-

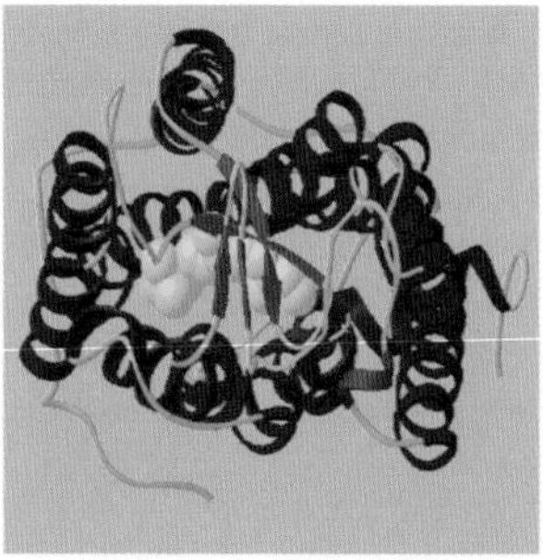

Fig. 1. The structure of bovine rhodopsin seen from "above"

tion by helix IV–helix IV contacts would allow for a force on the loop IV–V that might regulate the ligand entry in the other partner of the dimer. The rhodopsin crystal dimer structures do not resolve this question, as the experimental structures show antiparallel helix bundle pairs, whereas the natural dimers must be parallel bundle pairs. The β-hairpin between helices IV and V prevents access from the outside. This hairpin lies entirely between the helices, roughly parallel to the membrane surface. It has contacts with side chains of most of the helices. The most prominent contact is a disulphide bridge (Cys315–Cys480) to helix III. This calls for an explanation as to how ligands enter the binding cavity. For lipophilic ligands, like retinal itself (Schadel et al. 2003), entry/exit is expected to proceed via the membrane, as lipophilic ligands will accumulate in the membrane. For hydrophilic ligands, which include some peptides, insertion of the ligand will require some rearrangement of the loops including the hairpins. Not only will the hairpins have to make some adjustments, but the TMs will also move relative to one another. A number of clues as to what changes are likely to take place in the transition between the active and inactive structures have been published (Gouldson et al. 2004). In that work, and in a number of experimental studies cited therein (Gether and Kobilka 1998; Javitz et al. 1998), there is a movement of TM6 relative to TM3 and TM5. The crystal structure of bovine rhodopsin showed (Teller et al. 2001) that TM6 is unique in having only one hydrogen bond to another TM (TM7), while the other TMs are anchored by three or more interhelical hydrogen bonds.

2 Methods

Much GPCR-related research relies on access to all available data in a single easy-to-use data system, the GPCRDB. The principal data types contained in the GPCRDB are sequences, mutations, and structural information. Other GPCR-related information is accessible from the database's home page. Here we will describe the main steps of the GPCRDB update, its contents, and some of its functionalities.

The GPCRDB update procedure is handled by a series of python scripts, a MySQL database, and the WHAT IF (Vriend 1990) software.

Only the classification of new proteins having remote sequence similarity with already classified GPCRs, the definition of new families, and data checking in general, require some manual intervention and expertise. The other steps are fully automated.

GPCR proteins are imported from the Uniprot server (Bairoch et al. 2005). Receptors are then classified into the defined classes, families, and subfamilies using a profile-based method implemented in WHAT IF. Sequences that failed in the automatic classification step are further examined and classified manually. Fragments and short isoforms are put aside and are not used in the alignments in order to offer the highest possible alignment quality. For each class, family, and subfamily, WHAT IF is used to build multiple sequence alignments, phylogenetic trees, and other sequence-derived data in an automated manner. The profiles used for the alignments contain the location of the transmembrane domains and therefore allow us to ensure that the most conserved regions of the receptors are aligned without insertions and deletions. WHAT IF also produced the HTML pages to access the family-specific sequence data. cDNAs are imported from the EMBL databank (Cochrane et al. 2006) and aligned to their corresponding proteins using the genewise (Birney et al. 2004) program. Mutation data are identified and extracted from full-text articles with the MuteXt software (Horn et al. 2004). The latter automatically retrieves the corresponding UniProt entries, validates point mutations using sequence data and text mining approaches, and builds HTML pages to display mutation data as a function of receptors, articles, or residue positions. Multiple sequence alignments and snake-like diagrams (Campagne et al. 2003) are used to combine sequence, secondary structure, and mutation data. The use of the GPCRDB residue numbering system (Oliveira et al. 1993) permits this combination of many heterogeneous data types. A number is attributed to each residue in the seven transmembrane domains for all GPCR classes. This numbering system allows for fast comparisons between cognate residues in different receptors. In the mutation section of the database, the numbering system defined by Ballesteros and Weinstein (1995) is also indicated. Tables of available cross-references are provided for each GPCRDB entry to list all the different local information available and to ease navigation toward remote databases. The cross-references have been extracted from the UniProt entries and other databases. Among other data query and retrieval tools, a Blast search against the GPCRDB can be performed via the CMBI server. The GPCRDB data content is available via anonymous FTP from ftp://ftp.gpcr.org/pub/7tm/. A complete copy of the whole GPCRDB can be obtained upon request.

Copious amounts of data and speculative hypotheses can be found at the GPCRDB and they will not be reproduced here. The GPCRDB also contains a necropolis of earlier attempts at constructing GPCR models, and, more auspiciously for the future, assuming a steady accretion of good template structures, a detailed recipe for building models. Bovine rhodopsin and bacteriorhodopsin (Henderson and Schertler 1990; Pebay-Peyroula et al. 1997; Luecke et al. 1998) are sufficiently differently organized to make any detailed structural comparison meaningless (Unger et al. 1995, 1997; Schertler et al. 1993; Teller et al. 2001). However, in order to evaluate the quality of models based on the bacteriorhodopsin template, this superposition must be made. We therefore did this structure superposition by hand. Our recipe for determining the quality of bacteriorhodopsin-based BC models is given in Box 3.

Box 3 Recipe for judging BC model quality

Extract from the GPCRDB the alignment of the sequence of the GPCR model with the sequence of bovine rhodopsin.
Use the superposed structures to align the bovine rhodopsin sequence onto the bacteriorhodopsin sequence.
Extract from the modeling article how the authors aligned their GPCR with bacteriorhodopsin. If this alignment is not given, it can be extracted from a superposition of the bacteriorhodopsin-based GPCR model on the real bacteriorhodopsin structure.

This produces the alignment used for modeling. A comparison of the optimal alignment with the alignment used by the modeler is a good indication of the model's quality. This same method is used by the CASP competition judges to evaluate threading results (Venclovas et al. 2001). Our recipe for obtaining these BC-model alignment shifts differs from what is normally used because only the structure of bovine rhodopsin is known, while the beta-adrenergic receptor, for example, is the most modeled GPCR.

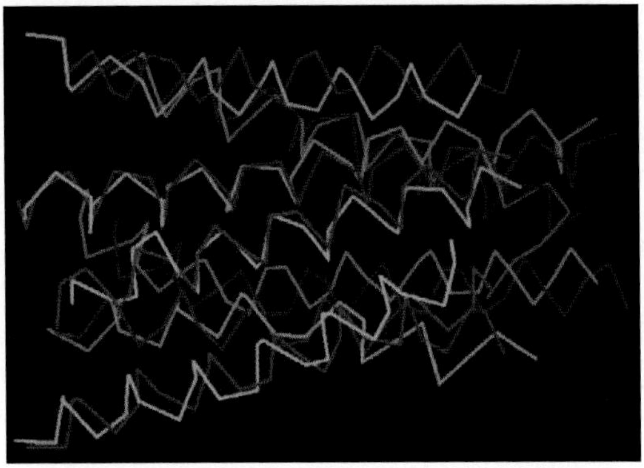

Fig. 2. Superposed bovine rhodopsin structure and BC-model

3 Results and Discussion

3.1 The Quality of BC Models

Figure 2 shows the superposition of the structure (Palczewski et al. 2000) and a very good BC model built, published (Oliveira et al. 1999), and deposited before August 2000. It can be seen that the gross features are modeled reasonably well. The Cα- and all-atom modeling errors (i.e., displacements between the model and the X-ray structure) are 2.5 Å and 3.2 Å, respectively. Although impressive, this model is still too poor to be of any use for rational drug design purposes.

The bovine rhodopsin structure (in red) is shown superposed on the BC model (in green) built and deposited before August 2000. As only the helices were modeled, the loops in the structure are also not shown.

We selected a superposition with a large overlap of the two retinal molecules. A shift in the structure superposition leads to a shift of three or four positions in the sequence alignment. Shifting the structure superposition up or down by one entire helical turn does not improve the alignments. Therefore, the subjective nature of the superposition does not influence our conclusions. We believe that all GPCR models (including our own) that are based on the bacteriorhodopsin template are poor, and none can have made a positive contribution to rational drug design projects, other than that common knowledge was confirmed (which is not surprising, as the models were first made to agree with that common knowledge).

Sequence alignments extracted from deposited GPCR models revealed that we could publish models that had residues misaligned by as many as ten positions. We think this holds a warning for the future.

3.2 How Could This Happen?

An extensive discussion of BC models can be found in the article section of the GPCRDB (Horn et al. 1998). None of the BC modelers had located the IV–V hairpin correctly between the helices. While all were aware of the Cys315–Cys480 disulphide bridge, which firmly anchors this loop near the top of TM III, all modelers 'knew' that loop IV–V had to be external. Often bizarre reasoning was used to reconcile these two contradicting claims and to justify the position of helix III. The experimental data enabling the correct prediction of the IV–V hairpin location was all the while available to the BC modelers. It also was known that in opsins His474 and Lys477 in this hairpin form a chloride-binding site that regulates the optimal absorption wavelength of the retinal (Wang et al. 1993). A reasonable conclusion from this is that since this site modifies the wavelength, it should be located near the retinal. Unfortunately, the common knowledge that the loops stick out into the solvent overcame the experimental and *in silico* (Kuipers et al. 1996) data about the chloride site. This provides a strong lesson for the future: models must explain all available data. If certain data seem untrustworthy, think twice. Most likely you do not trust those data only because it disagrees with your model.

Another problem that seriously hampered the quality of BC models is the massive irregularities in the transmembrane helices. Figure 3 shows some individual transmembrane helices. Helix II, for example, contains an α-bulge, i.e., one residue pair has a hydrogen-bonding pattern as if they are in a so-called π-helix. It is by no means certain that this is reproduced in other GPCRs; it may well be rhodopsin-specific (Bywater 2005).

The bovine rhodopsin structures provided us with a large number of structural surprises. Common knowledge had it that π helices (if it

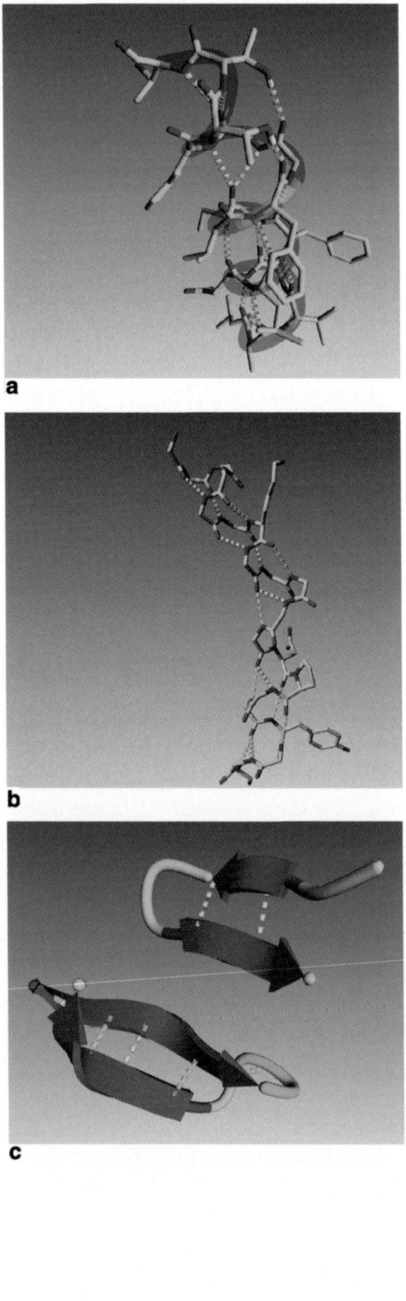
a
b
c

Fig. 3a–c. Irregularities in bovine rhodopsin helices. **a** The α-bulge in helix II. It can be clearly seen that the backbone C=O oxygen of the leucine located one turn after the aspartic acid of the conserved, well-known, LXXXD motif, forms hydrogen bonds with two backbone N–H protons, while all other backbone hydrogen bond donors and acceptors are satisfied. **b** The 3_{10} helix in the middle of helix VII. The lysine at the *top right* is the lysine that binds the retinal. The backbone C=O of the residue one turn below this lysine forms hydrogen bonds with two backbone N–H protons, while three backbone C=O groups and one backbone N–H group are not involved in hydrogen bonds. **c** The two β-hairpins near the top (extracellular side) of the structure. The two β-hairpins in the N-terminal arm and the loop IV–V are just 1 Å too far away from each other to form one contiguous sheet. It is unclear how uncommon such a separation is, but it certainly came very unexpectedly to the modeling community

involves one residue only, we had better call it an α bulge) are rare. In the excellent course "Principles of Protein Structure Using the Internet" (Johansson 1999) we find an extensive explanation of why this should be a rare event (see Box 4).

Box 4 Why the π helix (α bulge) is rare (Johansson 2006)

The π helix is an extremely rare secondary structural element in proteins. Hydrogen bonds within a π helix display a repeating pattern in which the backbone C=O of residue i hydrogen bonds to the backbone HN of residue i+5. Like the 3_{10} helix, one turn of the πhelix is sometimes found at the ends of regular α helices but π helices longer than a few i, i+5 hydrogen bonds are not found. The infrequency of this particular form of secondary structure stems from the following properties:

The Φ and Ψ angles of the pure π helix (–57.1, –69.7) lie at the very edge of an allowed, minimum energy region of the Ramachandran (Φ,Ψ) map.

The π helix requires that the angle τ (N–CA–C′) be larger (114.9) than the standard tetrahedral angle of 109.5°.

The large radius of the π helix means the polypeptide backbone is no longer in van der Waals contact across the helical axis forming an axial hole too small for solvent water to fill.

Side chains are more staggered than the ideal 3_{10} helix but not as well as the α helix.

Obviously there are big differences between an α bulge and a π helix, but while predicting the unpredictable, such fine details are easily overseen.

3.3 The Quality of AD Models

We were surprised to find many modeling studies performed after the release of the bovine rhodopsin three-dimensional coordinates into which very little knowledge of this template was incorporated. Ballesteros et al. (2001) wrote that amine receptors can be modeled from the bovine rhodopsin template. They neglect the IV–V hairpin, crystal contacts, and the fact that many residues cannot be detected in the X-ray structure. Orry and Wallace (2000) docked endothelin in an endothelin receptor model based on a rhodopsin model by Pogozheva (Pogozheva et al. 1997). The authors write in a note added after submission that the bovine rhodopsin structure became available after the paper was submitted, and claim that their model and the bovine rhodopsin structure are similar. Their model is not deposited, but from the figures in the article, it can be seen that the endothelin molecule is docked where one would expect the IV–V β hairpin and that this hairpin is modeled as a hyperexposed loop. These are just two of the many examples of neglect of details of the bovine rhodopsin structure. A survey of recent GPCR modeling-related literature revealed a series of flaws (see Box 5).

Box 5 Flaws detected in AD models

Total neglect of loops, especially the IV–V β-hairpin (Lopez-Rodriguez et al. 2001, 2002; Shim et al. 2003).

Modeling loop-based data for individual loops obtained from NMR experiments or from sequence similarity with another PDB file (Lequin et al. 2002; Chung et al. 2002; Yang et al. 2003; Mehler et al. 2002).

Molecular dynamics (MD) compacted the IV–V β hairpin (Pellegrini et al. 2001).

Models based on a frog electron density map (Church et al. 2002). It is regrettable that an MD publication on a homology model can be accepted for publication when the author has failed to show what the same protocol does to the bovine rhodopsin structure.

3.4 AD Modeling

The availability of the bovine rhodopsin structure opens new alleys for modeling GPCRs. Box 6, however, lists some warnings for would-be AD modelers.

Box 6 Warnings for Would-Be AD Modelers

The bovine rhodopsin structure is the inactive form of the protein, while the active form often is a much more appropriate modeling goal for pharmaceutical purposes.
The rhodopsin crystal structure is an antiparallel dimer, whereas GPCR dimers must be parallel.
It is far from certain that the bovine rhodopsin structure can be used as a template for all GPCRs, because sequence analyses indicates that opsins differ very much from the pharmaceutically interesting (Class A) GPCRs.
Modeling studies start with a sequence alignment between the bovine rhodopsin template and the GPCR model sequence. The percentage sequence identity between bovine rhodopsin and other (Class A) GPCRs can be as low as 20%.
Normally, when the sequence identity between the model and the template falls below 30%, the sequence alignment is the main bottleneck in the homology modeling procedure. Class A GPCRs might be an exception to this rule, because each helix contains one or two highly conserved residues that allow for an unambiguous alignment.
The observed structure of many loops seems to be determined by crystal contacts.
It is difficult to model the loops by homology, because most cytosolic loops cannot be seen in an electron density map, and most observed extracellular loop structures are probably induced by crystal packing forces. In any case, the sequence identity between most GPCRs and bovine rhodopsin is too low to derive any reliable loop alignment.
Several of the irregularities in the rhodopsin transmembrane helices seem rhodopsin-specific, whereas others seem more generic. It is not clear how to unambiguously decide which irregularities can be carried over from the rhodopsin template to, for example, an amine receptor model.

The bovine rhodopsin structure, combined with extensive sequence analyses (Horn et al. 1998), science philosophy, and all what we learned from the above, however, provides a series of hints (see Box 7).

Box 7 Hints for Would-Be AD Modelers

At three locations features can be seen that give hope for modeling. These are the highly conserved:

- Trp280 and Gly295 in loop II–III.
- Loop IV–V and the Cys315–Cys480 disulphide bridge.
- Tyr734 at the bend between the helices VII and VIII and the adjacent sequence motif Phe800, Arg/Lys801 in helix VIII.
- The associated WWW pages (articles section of the GPCRDB) lists many special positions.

3.5 The Active Form

Modeling the active form of Class A GPCRs depends critically on the hypothesized mechanism of that activation process (Gouldson et al. 2004). We therefore start with a summary of possible activation mechanisms. These activation models consist of essentially the same three general steps that are shown in Box 8.

Box 8 The three steps of the activation process

Entry of the ligand into the ligand binding pocket.
The receptor moving from the inactive state into the active state, or the active state being frozen by the ligand.
The G-protein being activated, or the activated state being frozen by the G-protein.

The clearest lesson to be learned from the BC experience is that molecular dynamics technology has not reached a level of maturity needed to aid with the prediction of the differences between the active and the inactive state. Not only is there a need for improved force fields that *inter alia* should reflect the membrane environment, but there are only few attempts to model the solvent (lipid bilayer with water above and below it) and none of these take account of, for example, the fact that the two leaflets of the bilayer are mutually asymmetric in character.

The low-resolution structure of light-activated bovine rhodopsin suggests that an outward motion of helices V and VI might be the major difference relative to the dark state (Szundi et al. 2006). Unfortunately, Murphy took care that those helices V and VI, in Schertler's crystal form (Schertler 2005), are involved in crystal contacts so that definitive conclusions about this motion cannot be drawn. The idea that especially the cytosolic half of helix VI moves a great deal upon GPCR activation would be in agreement with the results of our sequence analyses.

3.6 New Rules to Replace the Old Dogmas

Even if it may be deemed desirable to model loops (e.g., for studying the interactions between cytosolic loops and G-proteins, or between external loops and large peptide ligands) we would advise against this. There are no accurate template structures and precious little homology. The work by Yeagle et al. (1997, 1995, 2000) makes clear that determination of the structure of the loops independently from the rest of the molecule is not successful. Paradoxically, it may seem, it is harder to model oligopeptides than folded proteins, and this is because the former have no tertiary structure, and secondary structure is either absent or hard to predict.

Nevertheless, for most purposes (e.g., ligand design), it will be enough to model the seven transmembrane helices and the IV–V hairpin. The alignment of the helices should be based on the conserved motifs. Extrapolating from the performance of GPCR modelers over the years, we can only advise sticking to the bovine rhodopsin helix backbone coordinates. Any attempt to improve this for other GPCRs will undoubtedly make things worse rather than better. This unfortunately contradicts the earlier remarks that some of the helix irregularities might be rhodopsin-specific (Bywater 2005). The IV–V hairpin should be modeled from bovine rhodopsin. If this loop is not present in the model sequence, it seems doubtful that a reliable model structure can be built.

The bovine rhodopsin three-dimensional coordinates represent the inactive form of this receptor. To model the (pharmaceutically much more interesting) active form of GPCRs, one should not rely on molecular dynamics, but rather on the outcome of experiments that can be interpreted unambiguously (Gouldson et al. 2004). Molecular dynamics simulations might fill in the details once the relevant motions have been determined experimentally.

Modeling GPCRs based on their homology to rhodopsin is seriously hampered by the fact that we cannot predict well if the irregularities in the rhodopsin helices are rhodopsin-specific, or general GPCR or Class A GPCR features. Modeling is further hampered by the low level of identity between the sequences of the rhodopsin template and the other (Class A) receptors. However, it is commonly observed in homology modeling projects that the areas in and around active sites and other binding motifs are better conserved than the rest of the protein, and it can safely be assumed that this will also be the case for GPCRs. In the Class A GPCRs, the so-called rhodopsin-like family, the active site (G-protein binding) and the regulatory site (ligand binding) are both located in the transmembrane helix bundle. Consequently, we expect the conserved residues and the residues that are conserved at the subfamily level to reside in these transmembrane helices. Modeling only the bundle of seven helices, loop IV–V, and perhaps helix VIII should therefore suffice to arrive at a model that, although certainly wrong in most of its details, will at least be useful for designing experiments, especially mutagenesis experiments, which provide valuable feedback in terms of structure.

Experimental data clearly show the function of several residues. For example, the arginine at position 340 is the crucial switch in G-protein coupling, the tyrosine at position 528 is involved in G-protein binding, etc. Having coordinates available for just one GPCR, we have to resort to sequence-based techniques to obtain information about the location and role of residues for which the experimental function determination is less trivial. Correlated mutation analyses [CMA (Oliveira et al. 1993)] and entropy variability analyses [EVA (Oliveira et al. 2003a,b)] have revealed several types of residue positions that show recognizable conservation/variability patterns. Obviously, knowledge about these conservation patterns aids in the alignment of GPCR sequences to the rhodopsin structure. These conservation patterns range from very conserved in and around the active site where they have a role in maintaining the right structure and mobility required for the function, up to near maximal variability at positions where only very weak evolutionary pressures work in only a small subset of the sequences.

In practice, the most conserved residues form the active site, which for GPCRs is the G-protein binding site. Residues known to be involved in G-protein binding must be aligned in the multiple sequence alignment. The residues involved in maintaining the structure of the active site are less conserved than the active site residues themselves, but are still much more conserved than all other residues. The next group of yet lesser conserved residues are the core residues that form critical contacts between secondary structure elements. At these positions, one tends to observe mainly intermediately large residues that normally are hydrophobic, but in GPCRs also can be polar, or sometimes even charged. Several of these residues are involved in signal transduction through the receptor and show a corresponding EVA pattern. Occasionally, cysteines in a bridge are so conserved that automatic methods might see them as functionally important. But in general, residues that are only involved in maintaining the structural integrity of the molecule tend to be less conserved than active site residues. We therefore speculate that the cys–cys bridge between the external end of helix III and loop IV–V has a functional role, presumably in the regulation of ligand entry. Obviously, the residues facing the lipid membrane are in majority hydrophobic. Nonhydrophobic outward-pointing residues most likely have a functional role, perhaps in dimer formation.

We can draw the general conclusion that modeling is possible for all residue positions with recognizable conservation patterns, or with clearly recognizable variability patterns. Clearly, most opsins can be modeled from the rhodopsin template over nearly the full length. It also seems likely that the cytosolic halves of most Class A GPCRs can be modeled as well. It is at this moment still a matter of modeler's religion rather than exact science whether one models the whole helices or not. It is of crucial importance for pharmaceutical purposes to model the loop IV–V. This will be difficult in all cases where the loops have a different length from this loop in rhodopsin and in all cases where the sequence is very different from rhodopsin. Obviously, the cysteine in this loop IV–V must be aligned with the bridged cysteine in rhodopsin. Several sequences (e.g., melanocortin, mas, cannabinoid, and a few more) do not have this loop, which makes it highly unlikely that their helical organization will be anywhere similar to that of the rhodopsin helices. In a few cases, loop IV–V has several cysteines, which in many cases will require experimental determination of the right cysteine to align for the cys–cys bridge. In a few classes of GPCRs, a highly conserved proline is found at position 348. In case it is desired that helix VIII be part of the model, one can use the highly conserved motif F(Y,L)810-

R(K)811 that is present in rhodopsin and in several other GPCR classes.

So, in summary, we can say that all BC GPCR models were poor, and all AD GPCR models will not be up to CASP-competition (Protein Structure Prediction Center 2006) standards. We have given the beginning of a new GPCR modeling recipe. One day, hopefully soon (!) this recipe will be proven wrong, but it is the best we can do given current data and Ockham's razor. We can, however, draw some hope from the (paraphrased) quote of Bax that "All GPCR models are wrong, but sometimes these models can be useful." And when used with care, GPCR models are often a powerful tool to aid us with the design of experiments that can shed light on the sequence-structure-function-human-health relation of this intriguing class of molecules.

Acknowledgements. The GPCRDB was initiated as an EC sponsored project (PL 950224). GV acknowledges financial support from BioRange and BioSapiens. The BioSapiens project is funded by the European Commission within its FP6 Programme, under the thematic area "Life sciences, genomics and biotechnology for health," contract number LSHG-CT-2003–503265. BioRange is a programme of the Netherlands Bioinformatics Centre (NBIC), which is supported by a BSIK grant through the Netherlands Genomics Initiative (NGI).

References

Bairoch A, Apweiler R, Wu CH, Barker WC, Boeckmann B, Ferro S, Gasteiger E, Huang H, Lopez R, Magrane M, Martin MJ, Natale DA, O'Donovan C, Redaschi N, Yeh LS (2005) The universal protein resource (UniProt). Nucleic Acids Res 33:D154–D159

Ballesteros JA, Weinstein H (1995) Integrated methods for modeling G-protein coupled receptors. Methods Neurosci 25:366–428

Ballesteros JA, Shi L, Javitch JA (2001) Structural mimicry in G protein-coupled receptors: implications of the high-resolution structure of rhodopsin for structure–function analysis of rhodopsin-like receptors. Mol Pharmacol 60: 1–19

Beukers MW, Kristiansen I, Ijzerman AP, Edvardsen O (1999) TinyGRAP database: a bioinformatics tool to mine G protein-coupled receptor mutant data. TiPS 1999 20:475–477

Birney E, Clamp M, Durbin R (2004) GeneWise and Genomewise. Genome Res 14:988–995

Bramblett RD, Panu AM, Ballesteros JA, Reggio PH (1995) Construction of a 3D model of the cannabinoid CB1 receptor: determination of helix ends and helix orientation. Life Sci 56:1971–1982

Bywater RP (2005) Location and nature of the residues important for ligand recognition in G Protein-coupled receptors. J Mol Recognit 18:60–72

Campagne F, Bettler E, Vriend G, Weinstein H (2003) Batch mode generation of residue-based diagrams of proteins. Bioinformatics 19:1854–1855

Chung DA, Zuiderweg ER, Fowler CB, Soyer OS, Mosberg HI, Neubig RR (2002) NMR structure of the second intracellular loop of the alpha 2A adrenergic receptor: evidence for a novel cytoplasmic helix. Biochemistry 41:3596–3604

Church WB, Jones KA, Kuiper DA, Shine J, Iismaa TP (2002) Molecular modelling and site-directed mutagenesis of human GALR1 galanin receptor defines determinants of receptor subtype specificity. Protein Eng 5:313–323

Cochrane G, Adelbert P, Althorpe N et al. (2006) EMBL Nucleotide Sequence Database: developments in 2005. Nucleic Acids Res 34:D10–D15

Cronet P, Sander C, Vriend G (1993) Modelling of transmembrane seven helix bundles. Protein Eng 6:59–64

Donnelly D, Overington JP, Ruffle SV, Nugent JH, Blundell TL (1993) Modelling alpha-helical transmembrane domains: the calculation and use of substitution tables for lipid-facing residues. Protein Sci 2:55–70

Filizola M, Perez JJ, Carteni-Farina M (1998) BUNDLE: a program for building the transmembrane domains of G protein-coupled receptors. J Comput Aided Mol Des 12:111–118

Gether U, Kobilka BK (1998) G Protein receptor activation: II. Mechanism of agonist activation. J Biol Chem 273:17979–17982

Gouldson PR, Kidley NJ, Bywater RP, Psaroudakis G, Brooks HD, Diaz C, Shire D, Reynolds CA (2004) Toward the active conformations of rhodopsin and beta-2-adrenergic receptor. Proteins 56:67–84

Henderson R, Schertler GFX (1990) The structure of bacteriorhodopsin and its relevance to the visual opsins and other seven-helix G protein-coupled receptors. Philos Trans R Soc Lond B Biol Sci 326:379–389

Herzyk P, Hubbard RE (1998) Combined biophysical and biochemical information confirms arrangement of transmembrane helices visible from the three-dimensional map of frog rhodopsin. J Mol Biol 281:741–754

Horn F, Weare J, Beukers MW, Horsch S, Bairoch A, Chen W, Edvardsen O, Campagne F, Vriend G (1998) GPCRDB: an information system for G protein-coupled receptors. Nucleic Acids Res 26:275–279

Horn F, Bettler E, Oliveira L, Campagne F, Cohen FE, Vriend G (2003) GPCRDB information system for G protein-coupled receptors. Nucleic Acids Res 31:294–297

Horn F, Lau AL, Cohen FE (2004) Automated extraction of mutation data from the literature: application of MuteXt to G protein-coupled receptors and nuclear hormone receptors. Bioinformatics 20:557–568

Javitch JA, Ballesteros JA, Weinstein H, Chen J (1998) A cluster of aromatic residues in the sixth membrane-spanning segment of the D2 receptor is accessible to the binding-site crevice. Biochemistry 37:998–1006

Johansson K (1999) Bioinformatics practical. http://alpha2.bmc.uu.se/~kenth/bioinfo/structure/secondary/08.html. Cited 24 November 2006

Kuipers W, Van Wijngaarden I, Ijzerman AP (1994) A model of the serotonin 5-HT1A receptor: agonist and antagonist binding sites. Drug Des Discov 11:231–249

Kuipers W, Oliveira L, Paiva ACM, Rippmann F, Sander C, Vriend G, Ijzerman AP (1996) Sequence-function correlation in G protein-coupled receptors. In: Findlay JBC (ed) Membrane protein models. BIOS Scientific, Oxford

Kuipers W, Oliveira L, Vriend G, IJzerman AP (1997) Identification of class-determining residues in G protein-coupled receptors by sequence analysis. Receptors Channels 5:159–174

Lequin O, Bolbach G, Frank F, Convert O, Girault-Lagrange S, Chassaing G, Lavielle S, Sagan S (2002) Involvement of the second extracellular loop (E2) of the neurokinin-1 receptor in the binding of substance P. Photoaffinity labeling and modeling studies. J Biol Chem 277:22386–22394

Lopez-Rodriguez ML, Murcia M, Benhamu B, Olivella M, Campillo M, Pardo L (2001) Computational model of the complex between GR113808 and the 5-HT4 receptor guided by site-directed mutagenesis and the crystal structure of rhodopsin. J Comput Aided Mol Des 15:1025–1033

Lopez-Rodriguez ML, Vicente B, Deupi X, Barrondo S, Olivella M, Morcillo MJ, Behamu B, Ballesteros JA, Salles J, Pardo L (2002) Design, synthesis and pharmacological evaluation of 5-hydroxytryptamine(1a) receptor ligands to explore the three-dimensional structure of the receptor. Mol Pharmacol 62:15–21

Luecke H, Richter HT, Lanyi JK (1998) Proton transfer pathways in bacteriorhodopsin at 2.3 angstrom resolution. Science 280:1934–1937

Mehler EL, Periole X, Hassan SA, Weinstein H (2002) Key issues in the computational simulation of GPCR function: representation of loop domains. J Comput Aided Mol Des 16:841–853

Oliveira L, Paiva ACM, Vriend G (1993) A common motif in G protein-coupled seven transmembrane helix receptors. J Comput Aided Mol Des 7:649–658

Oliveira L, Paiva ACM, Vriend G (1999) A low resolution model for the interaction of G proteins with G protein-coupled receptors. Prot Eng 12:1087–1095

Oliveira L, Paiva PB, Paiva AC, Vriend G (2003a) Identification of functionally conserved residues with the use of entropy-variability plots. Proteins 52:544–552

Oliveira L, Paiva PB, Paiva AC, Vriend G (2003b) Sequence analysis reveals how G protein-coupled receptors transduce the signal to the G protein. Proteins 52:553–560

Orry AJW, Wallace BA (2000) Modelling and docking the endothelin G protein-coupled receptor. Biophys J 79:3083:3094

Palczewski K, Kumasaka T, Hori T, Behnke CA, Motoshima H, Fox BA, Le Trong I, Teller DC, Okada T, Stenkamp RE, Yamamoto M, Miyano M (2000) Crystal structure of rhodopsin: a G protein-coupled receptor. Science 289:739–745

Pardo L, Ballesteros JA, Osman R, Weinstein H (1992) On the use of the transmembrane domain of bacteriorhodopsin as a template for modeling the three-dimensional structure of guanine nucleotide-binding regulatory protein-coupled receptors. PNAS 89:4009-4012

Pebay-Peyroula E, Rummel G, Rosenbusch JP, Landau EM (1997) X-ray structure of bacteriorhodopsin at 2.5 angstroms from microcrystals grown in lipid cubic phases. Science 277:1676–1681

Pellegrini M, Bremer AA, Ulfers AL, Boyd ND, Mierke DF (2001) Molecular characterization of the substance P*neurokinin-1 receptor complex: development of an experimentally based model. J Biol Chem 276:22862–22867

Pogozheva ID, Lomize AL, Mosberg HI (1997) The transmembrane 7-alpha-bundle of rhodopsin: distance geometry calculations with hydrogen bonding constraints. Biophys J 72:1963–1985

Protein Structure Prediction Center (2006) http://predictioncenter.gc.ucdavis.edu/. Cited 24 November 2006

Prusis P, Schiöth HB, Muceniece R, Herzyk P, Afshar M, Hubbard RE, Wikberg JES (1997) Modelling of the three-dimensional structure of the human melanocortin 1 receptor, using an automated method and docking of a rigid cyclic melanocyte stimulating hormone core peptide. J Mol Graph Model 15:307–315

Rippmann F, Bottcher E (1993) Molecular modelling of serotonin receptors. 7TM 3:1–27

Schadel SA, Heck M, Maretzki D, Filipek S, Teller DC, Palczewski K, Hofmann KP (2003) Ligand channeling within a G-protein-coupled receptor. The entry and exit of retinals in native opsin. J Biol Chem 278:24896–24903

Schertler GF (2005) Structure of rhodopsin and the metarhodopsin I photointermediate. Curr Opin Struct Biol 15:408–415
Schertler GFX, Hargrave PA (1995) Projection structure of frog rhodopsin in two crystal forms. PNAS 192:11578–11582
Schertler GF, Villa C, Henderson R (1993) Projection structure of rhodopsin. Nature 362:770–772
Shi L, Javitch JA (2002) The binding site of aminergic G protein-coupled receptors: the transmembrane segments and second extracellular loop. Annu Rev Pharmacol Toxicol 42:437–467
Shim JY, Welsh WJ, Howlett AC (2003) Homology model of the CB1 cannabinoid receptor: sites critical for nonclassical cannabinoid agonist interaction. Biopolymers 71:169–189
Szundi I, Ruprecht JJ, Epps J, Villa C, Swartz TE, Lewis JW, Schertler GF, Kliger DS (2006) Rhodopsin photointermediates in two-dimensional crystals at physiological temperatures. Biochemistry 45:4974–4982
Takeda K, Sato H, Hino T, Kono M, Fukuda K, Sakurai I, Okada T, Kouyama T (1998) A novel three-dimensional crystal of bacteriorhodopsin obtained by successive fusion of the vesicular assemblies. J Mol Biol 283:463–474
Teller DC, Okada T, Behnke CA, Palczewski K, Stenkamp RE (2001) Advances in determination of a high-resolution three-dimensional structure of rhodopsin, a model of G protein-coupled receptors. Biochemistry 40:7761–7772
Unger VM, Schertler GFX (1995) Low resolution structure of bovine rhodopsin determined by electron cryo-microscopy. Biophys J 68:1776–1786
Unger VM, Hargrave PA, Baldwin JM, Schertler GFX (1997) Arrangement of rhodopsin transmembrane alpha-helices. Nature 389:203–206
Vaidehi N, Floriano WB, Trabanino R, Hall SE, Freddolino P, Choi EJ, Zamanakos G, Goddard WA 3rd (2002) Prediction of structure and function of G protein-coupled receptors. PNAS 2002 99:12622–12627
Venclovas C, Zemla A, Fidelis K, Moult J (2001) Comparison of performance in successive CASP experiments. Proteins Suppl 5:163–170
Vriend G (1990) WHAT IF: a molecular modeling and drug design program. J Mol Graph 8:52–56
Wang Z, Asenjo AB, Oprian DD (1993) Identification of the Cl-binding site in the human red and green colour vision pigments. Biochemistry 32:2125–2130
Watson S, Arkinstall S. The G-protein linked receptor Facts Book. 1994, Academic Press Ltd, ISBN 0-12-738440-5

Yang X, Wang Z, Dong W, Ling L, Yang H, Chen R (2003) Modeling and docking of the three-dimensional structure of the human melanocortin 4 receptor. J Protein Chem 22:335–344

Yeagle PL, Alderfer JL, Albert AD (1995) Structure of the third cytoplasmic loop of bovine rhodopsin. Biochemistry 34:14621–14625

Yeagle PL, Alderfer JL, Albert AD (1996) Structure determination of the fourth cytoplasmic loop and carboxyl terminal domain of bovine rhodopsin. Mol Vis 2:12–19

Yeagle PL, Alderfer JL, Salloum AC, Ali L, Albert AD (1997) The first and second cytoplasmic loops of the G protein-receptor, rhodopsin, independently form betaturns. Biochemistry 36:3864–3869

Yeagle PL, Salloum A, Chopra A, Bhawsar N, Ali L, Kuzmanovski G, Alderfer JL, Albert AD (2000) Structures of the intradiskal loops and amino terminus of the G-protein receptor, rhodopsin. J Pept Res 55:455–465

Ernst Schering Foundation Symposium Proceedings, Vol. 2, pp. 49–73
DOI 10.1007/2789_2006_003

Published Online: 16 May 2007

QSAR Modeling of GPCR Ligands: Methodologies and Examples of Applications

A. Tropsha(✉), S.X. Wang

The Laboratory for Molecular Modeling, CB#7360, Beard Hall, School of Pharmacy, University of North Carolina at Chapel Hill, 27599-7360 North Carolina, USA
email: *alex_tropsha@unc.edu*

Abstract. GPCR ligands represent not only one of the major classes of current drugs but the major continuing source of novel potent pharmaceutical agents. Because 3D structures of GPCRs as determined by experimental techniques are still unavailable, ligand-based drug discovery methods remain the major computational molecular modeling approaches to the analysis of growing data sets of tested GPCR ligands. This paper presents an overview of modern Quantitative Structure Activity Relationship (QSAR) modeling. We discuss the critical issue of model validation and the strategy for applying the successfully validated QSAR models to virtual screening of available chemical databases. We present several examples of applications of validated QSAR modeling approaches to GPCR ligands. We conclude with the comments on exciting developments in the QSAR modeling of GPCR ligands that focus on the study of emerging data sets of compounds with dual or even multiple activities against two or more of GPCRs.

1 Introduction

G-protein-coupled receptors represent the largest class of human proteins regulating vital biological and physiological functions. Naturally, these receptors have been regarded as major targets for drug discovery. Various estimates place the percentage of all modern drugs acting via GPCRs at 50%–70% (Flower 1999; Shay and Wright 2006). This large proportion of GPCR ligands among current drugs by no means implies that the GPCR drug discovery effort has exhausted itself. On the contrary, the advances in genomics and proteomics in recent years have led to identification of the growing number of novel GPCRs, many of which still have unknown physiological functions and are considered orphan receptors. Rapid growth of biomolecular databases of the ligands tested against panels of GPCRs such as PDSP Ki (Roth and Kroeze 2006) (cf. http://pdsp.med.unc.edu/) or GLIDA (Okuno et al. 2006) (cf. http://pharminfo.pharm.kyoto-u.ac.jp/services/glida/) emphasize the growing need in rationalizing the information about the relationships between structure and activity of all tested compounds. This challenge provides a natural avenue to explore computational molecular modeling and biomolecular informatics approaches to accelerate the focused discovery of novel potent GPCR ligands and ultimately realistic drug candidates even within the academic sector (Kozikowski et al. 2006).

Broadly speaking, computational drug discovery approaches include structure-based and ligand-based methods. The former require the knowledge of the three-dimensional structure of the target that can be obtained either using experimental approaches such as X-ray or nmr or based on protein homology modeling approaches. It is well known that even in those cases when the high-resolution X-ray structure of a target protein is available, accurate prediction of the bound poses of native ligands is a formidable challenge. Naturally, the use of receptor models for structure-based drug discovery studies using tools such as docking and scoring should be attempted with a great deal of caution. Nevertheless, since roughly 15 years ago (Hibert et al. 1991), great effort has gone into modeling various GPCRs and using such models to search for ligands that bind to the receptors (Bissantz et al. 2003).

Vriend was first to establish a database of 3D models for most of the known GPCRs (Horn et al. 1998) (currently available at http://www.gpcr.org/7tm/models/vriend/index.html), and recently Skolnick developed a new publicly available database of predicted GPCR structures (Zhang et al. 2006) (cf. http://cssb.biology.gatech.edu/skolnick/files/gpcr/gpcr.html). Notably, the GPCR database developed by Vriend comes with the following explicit disclaimer: "If you need models for any serious work, get help from a local modelling expert, and don't use these models"; unfortunately, this warning was frequently ignored by many users. Most of the time, structure-based ligand-binding studies are limited to demonstrating that receptor models are capable of accommodating known ligands in their binding sites. This is most certainly insufficient because rigorous modeling effort requires clear demonstration that the model does not identify false positives (i.e., does not favor binding of ligands that are known to be inactive) or false negatives, that it could discriminate between agonists and antagonists, explain mutagenesis data, and most importantly make accurate prediction of novel potent ligands. Nevertheless, there were several reports in the literature on successful design of novel active ligands using GPCR models (Becker et al. 2006; Vaidehi et al. 2006). Still, such successes are infrequent and structure-based design of GPCR ligands appears to be premature as a universal method of choice to address the problem.

On the other hand, ligand-based approaches rely directly on the available information about chemical structures of many tested ligands and their activity or binding constants. Careful analysis of the historic data with the goal of building statistically significant and rigorously validated models that relate compound chemical structure to its potency should aid experimental GPCR researchers to design novel specific ligands. Such analysis is a subject of the research method known as Quantitative Structure Activity Relationship (QSAR) modeling.

In this paper, we describe major tenets of predictive QSAR modeling followed by examples of application of this method to several data sets of GPCR ligands. We argue in the concluding remarks of this chapter that rapidly growing data sets of compound libraries tested against a panel of GPCRs provide unique opportunities for cheminformatics analysis of this data toward rational design of novel GPCR ligands.

2 Methodology

2.1 QSAR Modeling

The field of Quantitative Structure Activity Relationships (QSAR), as an integral part of computer-aided drug design and discovery, is experiencing one of the most exciting periods in its history. Modern QSAR approaches are characterized by the use of multiple descriptors of chemical structure combined with the application of both linear and nonlinear optimization approaches, and a strong emphasis on rigorous model validation to afford robust and predictive QSAR models. The most important recent developments in the field concur with a substantial increase in the size of experimental data sets available for the analysis and an increased application of QSAR models as virtual screening tools to discover biologically active molecules in chemical databases and/or virtual chemical libraries. The latter focus differs substantially from the traditional emphasis on developing so-called explanatory QSAR models characterized by high statistical significance but only as applied to training sets of molecules with known chemical structure and biological activity.

An inexperienced user or sometimes even an avid practitioner of QSAR could be easily confused by the diversity of methodologies and naming conventions used in QSAR studies. Two-dimensional (2D) or three-dimensional (3D) QSAR, variable selection or Artificial Neural Network methods, Comparative Molecular Field Analysis (CoMFA) or binary QSAR present examples of various terms that may appear to describe totally independent approaches, which cannot be generalized or even compared to each other. In fact, any QSAR method can be generally defined as an application of mathematical and statistical methods to the problem of finding empirical relationships (QSAR models) of the form

$$P_i = \hat{k}(D_1, D_2, \ldots D_n) \, ,$$

where P_i are biological activities (or other properties of interest) of molecules, D_1, D_2, ... D_n are calculated (or, sometimes, experimentally measured) structural properties (molecular descriptors) of compounds, and $\hat{k}$ is some empirically established mathematical transformation that should be applied to descriptors to calculate the property values for all molecules. The relationship between values of descriptors D and target properties P can be linear [(e.g., multiple linear regression (MLR) as in the Hansch QSAR approach], where target property can be predicted directly from the descriptor values or nonlinear (such as artificial neural networks or classification QSAR methods) where descriptor values are used in characterizing chemical similarity between molecules, which in turn is used to predict compound activity. In general, each compound can be represented by a point in a multidimensional space, in which descriptors D_1, D_2, ... D_n serve as independent coordinates of the compound. The goal of QSAR modeling is to establish a trend in the descriptor values, which parallels the trend in biological activity. All QSAR approaches imply, directly or indirectly, a simple similarity principle, which for a long time has provided a foundation for the experimental medicinal chemistry: compounds with similar structures are expected to have similar biological activities. This implies that points representing compounds with similar activities in multidimensional descriptor space should be geometrically close to each other, and vice versa.

Despite formal differences between various methodologies, any QSAR method is based on a QSAR table, which can be generalized as shown in Fig. 1. To initiate a QSAR study, this table must include some identifiers of chemical structures (e.g., company id numbers, first column of the table in Fig. 1), reliably measured values of biological activity (or any other target property of interest, e.g., solubility, metabolic transformation rate, etc., second column), and calculated values of molecular descriptors in all remaining columns (sometimes, experimentally determined physical properties of compounds could be used as descriptors as well).

The differences in various QSAR methodologies can be understood in terms of the types of *target property* values, *descriptors*, and *op-*

Structure Id	Target Property (EC_{50}, K_i, etc)	Structural properties (descriptors)			
Comp. 1	P1	D11	D12	...	D1n
Comp. 2	P2	D21	D22	...	D2n
...	...	"	"	"	"
Comp. m	Pm	Dm1	Dm2	...	Dmn

$$\{P\} = \hat{K}\{D\}$$

Fig. 1. Generalized QSAR table

timization algorithms used to relate descriptors to the target properties and generate statistically significant models. Target properties (regarded as dependent variables in statistical data modeling sense) can generally be of three types: *continuous* (i.e., real values covering certain range, e.g., IC50 values, or binding constants); *categorical-related* (e.g., classes of target properties covering certain range of values, e.g., active and inactive compounds, frequently encoded numerically for the purpose of the subsequent analysis as one (for active) or zero (for inactive), or adjacent classes of metabolic stability such as unstable, moderately stable, stable; and *categorical-unrelated* (i.e., classes of target properties that do not relate to each other in any continuum, e.g., compounds that belong to different pharmacological classes, or compounds that are classified as drugs vs non-drugs). As simple as it appears, understanding this classification is actually very important since the choice of descriptor types as well as modeling techniques is often dictated by the type of the target properties. Thus, in general the latter two types require classification modeling approaches, whereas the former type of the target properties allows using linear regression modeling. The corresponding methods of data analysis are referred to as classification or continuous property QSAR.

Chemical descriptors (or independent variables in terms of statistical data modeling) can be typically classified into two types: *continuous* (i.e., range of real values for example as simple as molecular weight or many molecular connectivity indices); or *categorical-related* (i.e., classes corresponding to adjacent ranges of real values, for example counts of functional groups or binary descriptors indicating presence or absence of a chemical functional group or an atom in a molecule). Descriptors can be generated from various representations of molecules, e.g., 2D chemical graphs or 3D molecular geometries, giving rise to the terms of 2D or 3D QSAR, respectively. Understanding these types of descriptors is also important for understanding basic principles of QSAR modeling since as stated above, any modeling implies establishing the correlation between chemical similarity between compounds and similarity between their target properties. Chemical similarity is calculated in the descriptor's space using various similarity metrics (see excellent reviews by Peter Willett on the subject, e.g., Downs and Willett 1996); thus the choice of the metric is dictated in many cases by the descriptor type. For instance, in case of continuous descriptor variables the Euclidean distance in descriptor space is a reasonable choice of the similarity metric, whereas in case of binary variables metrics such as the Tanimoto coefficient or Manhattan distance would appear more appropriate.

Finally, *correlation methods* (that can be used either with or without variable selection) can be classified into two major categories, i.e., *linear* [e.g., linear regression (LR), or principal component regression (PCR), or partial least squares (PLS)] or *nonlinear* [e.g., k nearest neighbor (kNN), recursive partitioning (RP), artificial neural networks (ANN), or support vector machines (SVM)]. Most QSAR researchers practice their preferred modeling techniques, and the choice of the technique is frequently coupled with the choice of descriptor types. However, there are recent attempts (discussed in more detail below) to combine various modeling techniques and descriptor types as applied to individual data sets.

In some cases, the types of biological data, the choice of descriptors, and the class of optimization methods are closely related and mutually inclusive. For instance, multiple linear regression can only be applied when a relatively small number of molecular descriptors are used (at least five or six times smaller than the total number of compounds) and the target property is characterized by a continuous range of values. The use of multiple descriptors makes it impossible to use MLR because of a high chance of spurious correlation (Topliss and Edwards 1979) and requires the use of PLS or nonlinear optimization techniques. However, in general, for any given data set a user could choose between various types of descriptors and various optimization schemes, combining them in a practically mix-and-match mode, to arrive at statistically significant QSAR models in a variety of ways. Thus in general, all QSAR models can be universally compared in terms of their statistical significance, and, most importantly, their ability to predict accurately biological activities (or other target properties) of molecules not included in the training set. This concept of the predictive ability as a universal characteristic of QSAR modeling independent of the particulars of individual approaches should be kept in mind as we consider examples of QSAR tool applications in the subsequent sections of this paper.

2.2 Approaches to Developing Validated and Predictive QSAR Models

The process of QSAR model development is typically divided into three steps: data preparation, data analysis, and model validation. The implementation of these steps is generally determined by the researchers' interests, experience, and software availability. The resulting models are then frequently employed, at least in theory, to design new molecules based on chemical features or trends found to be statistically significant with respect to underlying biological activity. For instance, the popular 3D QSAR approach CoMFA (Comparative Molecular Field Analysis) (Cramer et al. 1989) makes suggestions regarding steric or electronic modifications of the training set compounds that are likely to increase their activity.

The first stage includes the selection of a data set for QSAR studies and the calculation of molecular descriptors. The second stage deals with the selection of a statistical data analysis and correlation technique, either linear or nonlinear, such as Partial Least Squares (PLS) or Artificial Neural Networks (ANN). Many different algorithms and a variety of computer software are available for this purpose; in all approaches, descriptors serve as independent variables, and biological activities as dependent variables.

Typically, the final part of QSAR model development is model validation, (Golbraikh and Tropsha 2002a; Tropsha et al. 2003) when the true predictive power of the model is established. In essence, predictive power is one of the most important characteristics of QSAR models. It can be defined as the ability of the model to accurately predict the target property (e.g., biological activity) of compounds that were not used for model development. The typical problem of QSAR modeling is that at the time of model development a researcher only has, essentially, training set molecules, so predictive ability can only be characterized by statistical characteristics of the training set model, and not by true external validation.

Most of the QSAR modeling methods implement the leave-one-out (LOO) (or leave-some-out) cross-validation procedure. The outcome from this procedure is a cross-validated correlation coefficient q^2, which is calculated according to the following formula:

$$q^2 = 1 - \frac{\sum (y_i - \hat{y}_i)^2}{\sum (y_i - \bar{y})^2},$$

where y_i, $\hat{y}_i$, and $\bar{y}$ are the actual, estimated by LOO cross-validation procedure, and the average activities, respectively. The summations in (3.1) are performed over all compounds, which are used to build a model (training set). Frequently, q^2 is used as a criterion of both robustness and predictive ability of the model. Many authors consider high q^2 (for instance, $q^2 > 0.5$) an indicator or even the ultimate proof of the high predictive power of a QSAR model. They do not test the models for their ability to predict the activity of compounds of an external test set (i.e., compounds that have not been used in the QSAR model development). For instance, recent publications (Girones et al. 2000; Bordas et al. 2000; Fan et al. 2001; Suzuki et al. 2001) provide several examples where the authors claimed that their models had high predictive ability without even validating them with an external test set. Other authors validate their models using only one or two compounds that were not used in QSAR model development (Recanatini et al. 2000; Moron et al. 2000), and still claim that their models are highly predictive.

A widely used approach to establish model robustness is the randomization of response (Wold and Eriksson 1995) (i.e., in our case of activities). It consists of repeating the calculation procedure with randomized activities and subsequent probability assessment of the resultant statistics. Frequently, it is used along with the cross-validation. Sometimes models based on the randomized data have high q^2 values, which can be explained by a chance correlation or structural redundancy (Clark et al. 2001). If all QSAR models obtained in the Y-randomization test have relatively high both R^2 and LOO q^2, it implies that an acceptable QSAR model cannot be obtained for the given data set by the current modeling method.

Thus, it is still uncommon to test QSAR models (characterized by a reasonably high q^2) for their ability to accurately predict biological activities of compounds not included in the training set. In contrast with such expectations, it has been shown that if a test set with known values of biological activities is available for prediction, there exists no correlation between the LOO cross-validated q^2 and the correlation coefficient R^2 between the predicted and observed activities for the test set (Fig. 2). In our experience (Kubinyi et al. 1998; Golbraikh and Tropsha 2002a), this phenomenon is characteristic of many data sets and is independent of the descriptor types and optimization techniques used to develop training set models. In a recent review, we emphasized the importance of external validation in developing reliable models (Tropsha et al. 2003).

As was suggested in several recent publications both by our colleagues (Novellino et al. 1995; Norinder 1996; Zefirov and Palyulin 2001; Kubinyi et al. 1998) and us (Golbraikh and Tropsha 2002b; Tropsha et al. 2003), the only way to ensure the high predictive power of a QSAR model is to demonstrate a significant correlation between predicted and observed activities of compounds for a validation (test) set, which was not employed in model development. We have shown (Golbraikh and Tropsha 2002b; Tropsha et al. 2003) that various *commonly accepted* statistical characteristics of QSAR models derived for a training set are *insufficient* to establish and estimate the predictive power of QSAR models. We emphasize that external validation must be made, in fact, a mandatory part of model development. This goal can be achieved by a division of an experimental SAR data set into the training and test

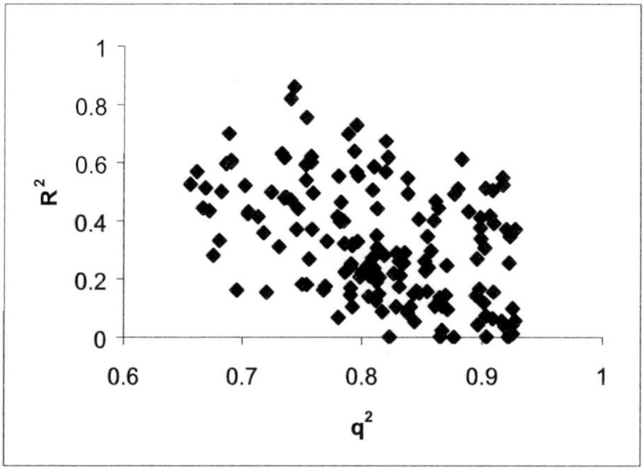

Fig. 2. Beware of q^2! External R^2 (for the test set) shows no correlation with the predictive LOO q^2 (for the training set). (Adapted from Golbraikh and Tropsha 2002a)

set, which are used for model development and validation, respectively. We believe that special approaches should be used to select a training set to ensure the highest significance and predictive power of QSAR models (Golbraikh and Tropsha 2002a; Golbraikh et al. 2003). Our recent reviews and publications describe several algorithms that can be employed for such division (Tropsha et al. 2003; Golbraikh and Tropsha 2002b; Golbraikh et al. 2003).

In order to estimate the true predictive power of a QSAR model, one needs to compare the predicted and observed activities of a sufficiently large external test set of compounds that were not used in the model development. One convenient parameter is an external q^2 defined as follows (similar to (3.1) for the training set):

$$q^2_{\text{ext}} = 1 - \frac{\sum_{i=1}^{\text{test}} \left(y_i - \hat{y}_i\right)^2}{\sum_{i=1}^{\text{test}} \left(y_i - \bar{y}_{\text{tr}}\right)^2}$$

where y_i and $\hat{y}_i$ are the measured and predicted (over the test set), respectively, values of the dependent variable and $\bar{y}_{\mathrm{tr}}$ is the averaged value of the dependent variable for the training set; the summations run over all compounds in the test set. Certainly, this formula is only meaningful when it does not differ significantly from the similar value for the test set (Oprea and Garcia 1996). In principle, given the entire collection of compounds with known structure and activity, there is no particular reason to select one particular group of compounds as a training (or test) set; thus, the division of the data set into multiple training and test sets (Golbraikh et al. 2003) or an interchangeable definition of these sets (Oprea 2001) is recommended.

The use of the following statistical characteristics of the test set was also recommended (Golbraikh et al. 2003): (i) correlation coefficient R^2 between the predicted and observed activities; (ii) coefficients of determination (predicted versus observed activities $R^2{}_0$); (iii) slopes k and k' of the regression lines through the origin. Thus, we consider a QSAR model predictive, if the following conditions are satisfied (Golbraikh et al. 2003):

$$q^2 > 0.5\,;$$

$$R^2 > 0.6\,;$$

$$\frac{\left(R^2 - R_0^2\right)}{R^2} < 0.1 \quad \text{or} \quad \frac{\left(R^2 - R_0'^2\right)}{R^2} < 0.1\,;$$

$$0.85 \le k \le 1.15 \quad \text{or} \quad 0.85 \le k' \le 1.15.$$

We have demonstrated (Golbraikh and Tropsha 2002b; Golbraikh et al. 2003) that all of the above criteria are indeed necessary to adequately assess the predictive ability of a QSAR model.

2.3 Combinatorial QSAR and a Workflow for Predictive QSAR Modeling

Our chief hypothesis is that if an implicit structure–activity relationship exists for a given data set, it can be formally manifested via a variety of QSAR models utilizing different descriptors and optimization protocols. We believe that multiple alternative QSAR models should be developed (as opposed to a single model using some favorite QSAR method) for each data set. Since QSAR modeling is relatively fast, these alternative models could be explored simultaneously when making predictions for external data sets. The consensus predictions of biological activity for novel test set compounds on the basis of several QSAR models, especially when they converge, are more reliable and provide better justification for the experimental exploration of hits. Finally, we emphasize the use of molecular descriptors as opposed to those based on chemical fragments, which affords exploration of a more diverse chemical space.

Our current approach to combi-QSAR modeling is summarized on the workflow diagram (Fig. 3). In most previously reported QSPR studies, the models are typically generated with a single modeling technique, frequently lacking external validation (Tropsha et al. 2002). Our experience suggests that QSPR is a highly experimental area of statistical data modeling where it is impossible to decide a priori which particular QSPR modeling method will prove to be most successful. Thus, to achieve QSPR models of the highest internal, and most importantly, *external* accuracy, we apply a combi-QSPR approach, which explores all possible combinations of various descriptor types and optimization methods along with external model validation. Each combination of descriptor sets and optimization techniques is likely to capture certain unique aspects of the structure–activity relationship. Our recent publications (Kovatcheva et al. 2004; de Cerqueira et al. 2006) demonstrate the power of the combi-QSAR approach in achieving significant models. Since our ultimate goal is to use the resulting models in compound property evaluation, application of different combinations of modeling techniques and descriptor sets will increase our chances for success. QSAR models are used increasingly in chemical data mining and combinatorial library design (Tropsha et al. 1999; Cho et al. 1998). For example, 3D stereoelectronic pharmacophore based on QSAR modeling was used recently to search the National Cancer Institute Repository of Small Molecules (2004) to find new leads for inhibiting HIV type 1 reverse transcriptase at the nonnucleoside binding site (Gussio et al. 1998). We recently introduced a descriptor pharmacophore concept (Tropsha and Zheng 2001) on the basis of variable selection QSAR: the descriptor pharmacophore is defined as a subset of molecu-

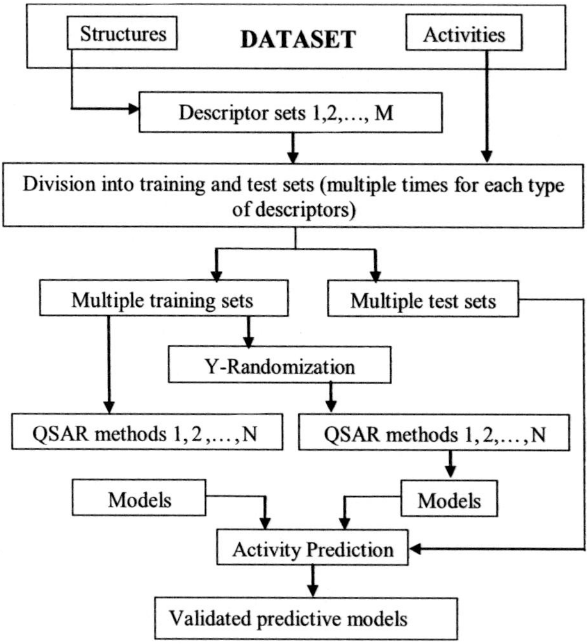

Fig. 3. Flow chart of the combinatorial QSAR methodology

lar descriptors that afford the most statistically significant QSAR model. It has been demonstrated that chemical similarity searches using descriptor pharmacophores as opposed to using all descriptors afford more efficient mining of chemical databases or virtual libraries to discover compounds with the desired biological activity (Tropsha and Zheng 2001; Shen et al. 2004).

The strategy for drug discovery that combines validated QSAR modeling and database mining has been under development in our group for several years (Shen et al. 2002; Tropsha et al. 2001) and was most recently applied successfully to the discovery of novel anticonvulsant agents (Shen et al. 2002) and more recently, to the analysis of D1 antagonists (Oloff et al. 2005). The approach is outlined in Fig. 4. It is important to stress that the outputs of these studies are not models

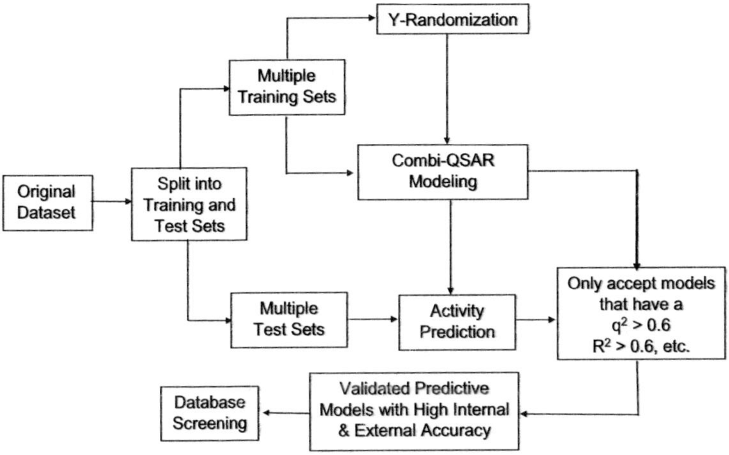

Fig. 4. Flowchart of predictive QSAR workflow based on validated combi-QSAR models

with their statistical characteristics as is typical for most QSAR studies. Rather, the modeling results are the predictions of the target properties for all database or virtual library compounds, which allows for immediate compound prioritization for subsequent experimental verification. Another advantage of using QSAR models for database mining is that this approach affords not only the identification of compounds of interest, but also quantitative prediction of the compounds' potency application to the discovery of novel potent compounds by means of database mining (Shen et al. 2004).

We shall emphasize that this approach shifts the traditional focus of QSAR modeling from achieving statistically significant training set models (where the results are presented in the form of statistical parameters) to identifying novel, potentially active compounds on the basis of statistically significant and externally validated models (i.e., where the results are presented in the form of compounds). We believe that this shift brings QSAR modeling in tune with the ultimate needs of experimental medicinal chemists in novel compounds rather than models. We use this approach routinely for the analyses of available experimental data sets, as illustrated by examples presented below.

3 Applications

3.1 QSAR Modeling of D_1 Dopaminergic Antagonists and Virtual Screening of Chemical Databases with Validated Models

Dopamine receptors are made of two subclasses [D_1-like and D_2-like subtypes (Kebabian and Calne 1979)] coded from five genes. The dopamine receptors play important roles such as modulation of motor function, cognition, memory, emotional activity, and various peripheral functions (Strange 1993) and have been especially implicated in disorders such as Parkinson's disease and schizophrenia (Seeman et al. 1987). The consequences of activation or blockade of dopamine receptors are wide-ranging (Creese and Iversen 1973; Phillips and Fibiger 1973; Pijnenburg et al. 1976; Ungerstedt and Arbuthnott 1970), and perturbation of dopamine neurotransmission may result in profound neurological, psychiatric, or physiological signs and symptoms. For these reasons, there has been a great deal of research focused on the discovery of novel dopaminergic ligands as potential drug candidates.

As is true for any GPCR, due to the difficulties with crystallizing transmembrane proteins, X-ray structures for the dopamine receptors are not currently available. This leaves ligand-based drug discovery approaches such as QSAR modeling the methods of choice for the analysis of known ligands and discovery of novel ligands for the dopamine receptors. Recently we have developed rigorously validated QSAR models for 48 D_1 dopaminergic antagonists and applied these models for virtual screening of available chemical databases to identify novel potent compounds (Oloff et al. 2005).

The pharmacological data for the 48 D_1 antagonists used in this study were reported elsewhere (Charifson et al. 1988, 1989; Minor et al. 1994; Schulz et al. 1984). Following our general strategy (Fig. 4), we applied the combinatorial QSAR modeling and model validation workflow to this data set. Thus, several QSAR methods were employed, including Comparative Molecular Field Analysis (CoMFA) (Cramer, III et al. 1988), Simulated Annealing-Partial Least Squares (SA-PLS) (Cho et al. 1998), k-Nearest Neighbor (kNN), (Zheng and Tropsha 2000), and Support Vector Machines (SVM) (Oloff et al. 2005). With the exception of

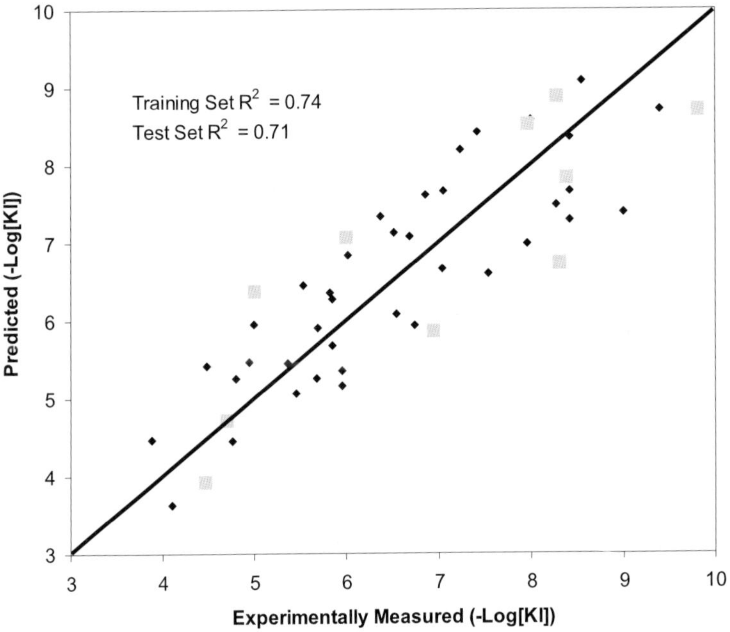

Fig. 5. Comparison of actual vs predicted D_1 antagonist binding affinity based on SVM QSAR models. The results are shown for both training (38 compounds; *dark squares*) and test (10 compounds, *gray squares*)

CoMFA, these approaches employed 2D topological descriptors generated with the MolConnZ software package (EduSoft 2003). The original data set was split into training and test sets to allow for external validation of each training set model. The resulting models were characterized by cross-validated R^2 (q^2) for the training set and predictive R^2 values for the test set of (q^2/R^2) 0.51/0.47 for CoMFA, 0.7/0.76 for kNN, R^2 for the training and test sets of 0.74/0.71 for SVM, and training set fitness and test set R^2 values of 0.68/0.63 for SA-PLS. Figure 5 illustrates a typical result of the modeling exercise.

Validated QSAR models with R^2 greater than 0.7 (for the training sets), (i.e., kNN and SVM) were used to mine three publicly available chemical databases: the National Cancer Institute (NCI) (NCI 2005) database of approximately 250,000 compounds, the Maybridge Database (Maybridge 2005) of approximately 56,000 compounds, and the ChemDiv Database containing approximately 450,000 compounds (Chemical Diversity 2004). These searches resulted in only 54 *consensus* hits (i.e., predicted active by all models); five of them were previously characterized as dopamine D_1 ligands, but were not present in the original data set. A small fraction of the purported D_1 ligands did not contain a catechol ring found in all known dopamine full agonist ligands, suggesting they may be novel structural antagonist leads. Although further experimental work is needed to verify this hypothesis, the identification of known ligands suggests that models produced in this study are capable of detecting novel D_1 compounds from large chemical databases.

3.2 Combinatorial QSAR Modeling of Serotonin Receptor 5HT1E/5HT1F Ligands and Subtype Selectivity

As mentioned in the introductory part of this paper, recent developments in high-throughput technologies afforded opportunities in the development of experimental screening programs where multiple ligands are tested simultaneously against panels of GPCRs (Roth and Kroeze 2006; Okuno et al. 2006). Such databases present new challenges for molecular modelers prompting the development of computational chemical genomics approaches. Unlike traditional QSAR modeling where we build models for compounds acting at single receptors, we are now poised to develop approaches that relate chemical structure to patterns of biological activity against two or more receptors. The first examples of such studies are beginning to appear in the literature (e.g., Blower et al. 2002), but we face the need for the systematic computational analysis of the receptorome, as termed by Roth and colleagues (Roth and Kroeze 2006). Our group has recently initiated such a research program and we present our initial results here as applied to the analysis of twp subtypes of 5HT receptor, i.e., 5HT1E and 5HT1F.

The PDSP Ki dataset (http://pdsp.med.unc.edu/) includes data on binding constants for 51 5HT1E ligands and 29 5HT1F ligands. We have employed the kNN modeling approach and started from developing individual QSAR models for each data set, followed by addressing the subtype specificity of the ligands acting at both receptors. Model building included multiple divisions of the original data set into training and test sets. The training and test set sizes varied from 46 and 5 to 28 and 23 compounds for 5HT1E data set and from 24 and 5 to 16 and 13 for 5HT1F data set, respectively. Generally, kNN models with leave-one-out cross-validated R^2 (q^2) values for the training set greater than 0.60, and linear fit predictive R^2 values for the external test set greater than 0.60 were accepted. The kNN method produced highly predictive models with q^2/R^2 values of 0.75/0.67 for 5HT1E ligands and 0.84/0.64 for 5HT1F ligands using MolConnZ descriptors. Similarly, models built using MOE descriptors generated statistical models with q^2/R^2 values of 0.61/0.75 for 5HT1E ligands and 0.61/0.68 for 5HT1F ligands. These results suggest that intrinsic structure–activity relationships exist for 5HT1E/5HT1F ligands exemplified by successful models developed with different types of molecular descriptors.

To quantify the selectivity of 5HT1E/5HT1F binders, a classification model has been developed to account for common binders (category 1) or non-common binders (category 0). The ratio of $Ki_{,5HT1E}$ to $Ki_{,5HT1F}$ was calculated as the selectivity index and the threshold was chosen to be $Ki_{,5HT1E}/Ki_{,5HT1F} = 3.0$ and 0.3. A higher threshold, such as $Ki_{,5HT1E}/Ki_{,5HT1F} = 1.0$ or 2.0, which others have used (Sutherland and Weaver 2004), resulted in too few common binders. Considering the errors of Ki values in the PDSP Ki database caused by broad assay sources, the difference in Ki values within tenfold is indistinguishable. Thus $Ki_{,5HT1E}/Ki_{,5HT1F} = 3.0$ and 0.3 have greater representation to the common/non-common category and showed the best classification statistics.

For the 5HT1E/5HT1F subtype selectivity data set, the training and test set sizes varied from 17 and 5 to 12 and 10, respectively. The accuracy of models was estimated using the Correct Classification Rate (CCR) defined as $N_{corr}/N = (TP + TN)/N$, where N and N_{corr} were the total number of compounds and the number of correctly classified compounds.

A high fraction of kNN QSAR models for the training sets were found to be statistically acceptable with $CCR_{train} \geq 0.70$ and many models achieved the impressive CCR = 1.00 for the validation sets. Usually models with $CCR_{test} \geq 0.70$ appeared to have $CCR_{train} \geq 0.70$, but the opposite was not always true. Thus, the models with high values of both CCR_{train} and CCR_{test} were considered acceptable. Remarkably, the kNN classification models predicted all seven common binders correctly out of 15 compounds in the training set and all six common binders out of seven compounds in the validation test set. We are currently investigating the difference in chemical patterns responsible for binding at individual receptor sites vs those responsible for selectivity. Virtual screening studies are also underway to identify novel putative selective binders.

4 Final Thoughts on QSAR Modeling

Our general approach to QSAR modeling goes beyond the traditional boundaries of this method. Although QSAR modeling is generally regarded as a ligand optimization approach that may lead to rational design of novel compounds, examples of rationally designed compounds are rare in any traditional QSAR modeling paper. Most of the publications present models that are capable of reproducing training set compound activity with high accuracy (in some cases, test set compound predictions are included, but those already have their biological activity determined). Thus, a typical outcome of a traditional QSAR modeling study is a set of statistical characteristics such as q^2, R^2, F value, etc., mostly for the training set, which provide little help to chemists interested in the design of novel molecules (CoMFA presents a notable exception by formally providing structural design hypothesis based on "fields").

The approach described in this paper does not stop when one could obtain a statistically significant training set model. Our approach places the emphasis of the *entire* QSAR modeling study on making reliable predictions of *chemical structures* expected to have the desired biological activity, rather than on respectable statistical characteristics of (training set) models. These predicted structures are either already available in existing chemical databases or are synthetically feasible (i.e., included in virtual combinatorial chemical libraries, which can also be mined with QSAR models). We believe that this extended view of the entire QSAR modeling approach exemplified by our recent studies of anticonvulsants (Shen et al. 2004) and D_1 antagonists (Oloff et al. 2005) presented in this paper brings the focus of modeling closer to the needs of medicinal chemists who both supply computational chemists with experimental structure–activity data and expect novel structures rather than equations and statistical parameters in return. We suggest that our approach that combines predictive QSAR modeling and database mining provides an important general avenue toward GPCR drug discovery that can be explored for many pharmacological data sets.

Acknowledgements. This work was supported in part by the NIH research grant GM066940 and by Berlex Biosciences. We appreciate fruitful discussions with Drs. R. Horuk and Sabine Schlyer.

References

Becker OM, Dhanoa DS, Marantz Y, Chen D, Shacham S, Cheruku S, Heifetz A, Mohanty P, Fichman M, Sharadendu A, Nudelman R, Kauffman M, Noiman S (2006) An integrated in silico 3D model-driven discovery of a novel, potent, and selective amidosulfonamide 5-HT1A agonist (PRX-00023) for the treatment of anxiety and depression. J Med Chem 49: 3116–3135

Bissantz C, Bernard P, Hibert M, Rognan D (2003) Protein-based virtual screening of chemical databases. II. Are homology models of G-protein coupled receptors suitable targets? Proteins 50:5–25

Blower PE, Yang C, Fligner MA, Verducci JS, Yu L, Richman S, Weinstein JN (2002) Pharmacogenomic analysis: correlating molecular substructure classes with microarray gene expression data. Pharmacogenomics J 2: 259–271

Bordas B, Komives T, Szanto Z, Lopata A (2000) Comparative three-dimensional quantitative structure-activity relationship study of safeners and herbicides. J Agric Food Chem 48:926–931

Charifson PS, Wyrick SD, Hoffman AJ, Simmons RM, Bowen JP, McDougald DL, Mailman RB (1988) Synthesis and pharmacological characterization of 1-phenyl-, 4-phenyl-, and 1-benzyl-1,2,3,4-tetrahydroisoquinolines as dopamine receptor ligands. J Med Chem 31:1941–1946

Charifson PS, Bowen JP, Wyrick SD, Hoffman AJ, Cory M, McPhail AT, Mailman R B (1989) Conformational analysis and molecular modeling of 1-phenyl-, 4-phenyl-, and 1-benzyl-1,2,3,4-tetrahydroisoquinolines as D1 dopamine receptor ligands. J Med Chem 32:2050–2058

Chemical Diversity (2004) ChemDiv Chemical Database. www.chemdiv.com. Cited 28 November 2006

Cho SJ, Zheng W, Tropsha A (1998) Rational combinatorial library design. Rational design of targeted combinatorial peptide libraries using chemical similarity probe and the inverse QSAR approaches. J Chem Inf Comput Sci 38:259–268

Clark RD, Sprous DG, Leonard JM (2001) Validating models based on large dataset. In: Höltje H-D, Sippl W (eds) Rational approaches to drug design, Proceedings of the 13th European Symposium on Quantitative Structure-Activity Relationship, Aug 27–Sept 1. Prous Science, Düsseldorf, pp 475–485

Cramer RD III, Patterson DE, Bunce JD (1988) Comparative Molecular Field Analysis (CoMFA) Effect of Shape on Binding of Steroids to Carrier Proteins. J Am Chem Soc 110:5959–5967

Cramer RD III, Patterson DE, Bunce JD (1989) Recent advances in comparative molecular field analysis (CoMFA). Prog Clin Biol Res 291:161–165

Creese I, Iversen SD (1973) Blockage of amphetamine induced motor stimulation and stereotypy in the adult rat following neonatal treatment with 6-hydroxydopamine. Brain Res 55:369–382

De Cerqueira LP, Golbraikh A, Oloff S, Xiao Y, Tropsha A (2006) Combinatorial QSAR modeling of P-glycoprotein substrates. J Chem Inf Model 46:1245–1254

Downs GM, Willett P (1996) Similarity searching in databases of chemical structures. In: Lipkowitz KB, Boyd D (eds) Reviews in computational chemistry. VCH Publishers, New York, pp 1–65

EduSoft L (2003) MolconnZ version 4.05. http://www.eslc.vabiotech.com/ [4.05]

Fan Y, Shi LM, Kohn KW, Pommier Y, Weinstein JN (2001) Quantitative structure-antitumor activity relationships of camptothecin analogues: cluster analysis and genetic algorithm-based studies. J Med Chem 44:3254–3263

Flower DR (1999) Modelling G-protein-coupled receptors for drug design. Biochim Biophys Acta 1422:207–234

Girones X, Gallegos A, Carbo-Dorca R (2000) Modeling antimalarial activity: application of kinetic energy density quantum similarity measures as descriptors in QSAR. J Chem Inf Comput Sci 40:1400–1407

Golbraikh A, Tropsha A (2002a) Beware of q^2! J Mol Graph Model 20:269–276

Golbraikh A, Tropsha A (2002b) Predictive QSAR modeling based on diversity sampling of experimental datasets for the training and test set selection. J Comput Aided Mol Des 16:357–369

Golbraikh A, Shen M, Xiao Z, Xiao YD, Lee KH, Tropsha A (2003) Rational selection of training and test sets for the development of validated QSAR models. J Comput Aided Mol Des 17:241–253

Gussio R, Pattabiraman N, Kellogg GE, Zaharevitz DW (1998) Use of 3D QSAR methodology for data mining the National Cancer Institute Repository of Small Molecules: application to HIV-1 reverse transcriptase inhibition. Methods 14:255–263

Hibert MF, Trumpp-Kallmeyer S, Bruinvels A, Hoflack J (1991) Three-dimensional models of neurotransmitter G-binding protein-coupled receptors. Mol Pharmacol 40:8–15

Horn F, Weare J, Beukers MW, Horsch S, Bairoch A, Chen W, Edvardsen O, Campagne F, Vriend G (1998) GPCRDB: an information system for G protein-coupled receptors. Nucleic Acids Res 26:275–279

Kebabian JW, Calne DB (1979) Multiple receptors for dopamine. Nature 277: 93–96

Kovatcheva A, Golbraikh A, Oloff S, Xiao YD, Zheng W, Wolschann P, Buchbauer G, Tropsha A (2004) Combinatorial QSAR of ambergris fragrance compounds. J Chem Inf Comput Sci 44:582–595

Kozikowski AP, Roth B, Tropsha A (2006) Why academic drug discovery makes sense. Science 313:1235–1236

Kubinyi H, Hamprecht FA, Mietzner T (1998) Three-dimensional quantitative similarity-activity relationships (3D QSiAR) from SEAL similarity matrices. J Med Chem 41:2553–2564

Maybridge (2005) http://www.daylight.com/products/databases/Maybridge html

Minor DL, Wyrick SD, Charifson PS, Watts VJ, Nichols DE, Mailman RB (1994) Synthesis and molecular modeling of 1-phenyl-1,2,3,4-tetrahydroisoquinolines and related 5,6,8,9-tetrahydro-13bH-dibenzo[a,h]quinolizines as D1 dopamine antagonists. J Med Chem 37:4317–4328

Moron JA, Campillo M, Perez V, Unzeta M, Pardo L (2000) Molecular determinants of MAO selectivity in a series of indolylmethylamine derivatives: biological activities, 3D-QSAR/CoMFA analysis, and computational simulation of ligand recognition. J Med Chem 43:1684–1691

National Cancer Institute (2004) Smiles strings. http://dtp.nci.nih.gov/docs/3d_database/structural_information/smiles_strings.html. Cited 28 November 2006

National Cancer Institute (2005) http://dtp.nci.nih.gov/docs/3d_database/structural_information/smiles_strings html

Norinder U (1996) Single and domain made variable selection in 3D QSAR applications. J Chemomet 10:95–105

Novellino E, Fattorusso C, Greco G (1995) Use of comparative molecular field analysis and cluster analysis in series design. Pharm Acta Helv 70:149–154

Okuno Y, Yang J, Taneishi K, Yabuuchi H, Tsujimoto G (2006) GLIDA: GPCR-ligand database for chemical genomic drug discovery. Nucleic Acids Res 34:D673–D677

Oloff S, Mailman RB, Tropsha A (2005) Application of validated QSAR models of D1 dopaminergic antagonists for database mining. J Med Chem 48: 7322–7332

Oprea TI (2001) Rapid estimation of hydrophobicity for virtual combinatorial library analysis. SAR QSAR Environ Res 12:129–141

Oprea TI, Garcia A E (1996) Three-dimensional quantitative structure-activity relationships of steroid aromatase inhibitors. J Comput Aided Mol Des 110:186–200

Phillips AG, Fibiger HC (1973) Dopaminergic and noradrenergic substrates of positive reinforcement: differential effects of d- and l-amphetamine. Science 179:575–577

Pijnenburg AJ, Honig WM, Van der Heyden JA, Van Rossum JM (1976) Effects of chemical stimulation of the mesolimbic dopamine system upon locomotor activity. Eur J Pharmacol 35:45–58

Recanatini M, Cavalli A, Belluti F, Piazzi L, Rampa A, Bisi A, Gobbi S, Valenti P, Andrisano V, Bartolini M, Cavrini V (2000) SAR of 9-amino-1,2,3,4-tetrahydroacridine-based acetylcholinesterase inhibitors: synthesis, enzyme inhibitory activity, QSAR, and structure-based CoMFA of tacrine analogues. J Med Chem 43:2007–2018

Roth BL, Kroeze WK (2006) Screening the receptorome yields validated molecular targets for drug discovery. Curr Pharm Des 12:1785–1795

Schulz DW, Wyrick SD, Mailman RB (1984) [3H]SCH23390 has the characteristics of a dopamine receptor ligand in the rat central nervous system. Eur J Pharmacol 106:211–212

Seeman P, Bzowej NH, Guan HC, Bergeron C, Reynolds GP, Bird ED, Riederer P, Jellinger K, Tourtellotte WW (1987) Human brain D_1 and D_2 dopamine receptors in schizophrenia, Alzheimer's, Parkinson's, and Huntington's diseases. Neuropsychopharmacology 1:5–15

Shay JW, Wright WE (2006) Telomerase therapeutics for cancer: challenges and new directions. Nat Rev Drug Discov 5:577–584

Shen M, LeTiran A, Xiao Y, Golbraikh A, Kohn H, Tropsha A (2002) Quantitative structure-activity relationship analysis of functionalized amino acid anticonvulsant agents using k nearest neighbor and simulated annealing PLS methods. J Med Chem 45:2811–2823

Shen M, Beguin C, Golbraikh A, Stables J, Kohn H, Tropsha A (2004) Application of predictive QSAR models to database mining: identification and experimental validation of novel anticonvulsant compounds. J Med Chem 47:2356–2364
Strange PG (1993) Brain biochemistry and brain disorders. Oxford University Press, New York
Sutherland JJ, Weaver DF (2004) Three-dimensional quantitative structure-activity and structure-selectivity relationships of dihydrofolate reductase inhibitors. J Comput Aided Mol Des 18:309–331
Suzuki T, Ide K, Ishida M, Shapiro S (2001) Classification of environmental estrogens by physicochemical properties using principal component analysis and hierarchical cluster analysis. J Chem Inf Comput Sci 41:718–726
Topliss JG, Edwards RP (1979) Chance factors in studies of quantitative structure-activity relationships. J Med Chem 22:1238–1244
Tropsha A, Zheng W (2001) Identification of the descriptor pharmacophores using variable selection QSAR: applications to database mining. Curr Pharm Des 7:599–612
Wold S, Eriksson L (1995) Statistical validation of QSAR results. In: Waterbeemd HVD (ed) Chemometrics methods in molecular design. VCH pp 309–318
Zefirov NS, Palyulin VA (2001) QSAR for boiling points of "small" sulfides. Are the "high-quality structure-property-activity regressions" the real high quality QSAR models? J Chem Inf Comput Sci 41:1022–1027
Zhang Y, Devries ME, Skolnick J (2006) Structure modeling of all identified G protein-coupled receptors in the human genome. PLoS Comput Biol 2:e13
Zheng W, Tropsha A (2000) Novel variable selection quantitative structure–property relationship approach based on the k-nearest-neighbor principle. J Chem Inf Comput Sci 40:185–194

Ernst Schering Foundation Symposium Proceedings, Vol. 2, pp. 75–91
DOI 10.1007/2789_2006_004

Published Online: 16 May 2007

Privileged Structures in GPCRs

R.P. Bywater(✉)

Magdalen College, OX1 4AU Oxford, England
email: *robert.bywater@magd.ox.ac.uk*

Systematic residue numbering convention:
GPCRs are of varying sizes, anywhere between approximately 315 and approximately 620 residues in length. This clearly poses problems for the creation of a standard numbering scheme for comparing residue positions across families. However, the length of the TM segments is more or less constant, and each TM has a pattern of conserved residues. Various schemes have been devised to pivot the numbering system about the most conserved residue in each TM. The GPCRDB system is just one of these. In this work, the residue numbers will take the form: (TM number)_(Position in that TM from the N-terminal end). Conversion to other standard numbering conventions can be made by consulting Bywater (2005) or Bondensgaard et al. (2004).

Orientation convention:
Unlike molecules of water-soluble proteins, which are oriented either in the frame of the user's 3D graphics device or with the orientation that they had in the crystal lattice, membrane proteins should be considered in relation to their orientation relative to the membrane, usually this is horizontal (in the x, y plane). In the convention employed here, the cytosolic side of the membrane is placed below the membrane bilayer and the extracellular side above. Directional prepositions used in the text, "up" and "down", "deeper down" etc. (i.e. along the z axis), are defined according to this convention.

Abstract. Certain kinds of ligand substructures recur frequently in pharmacologically successful synthetic compounds. For this reason they are called privileged structures. In seeking an explanation for this phenomenon, it is observed that the privileged structure represents a generic substructure that matches commonly recurring conserved structural motifs in the target proteins, which may otherwise be quite diverse in sequence and function. Using sequence-handling tools, it is possible to identify which other receptors may respond to the ligand, as dictated on the one hand by the nature of the privileged substructure itself or by the rest of the ligand in which a more specific message resides. It is suggested that privileged structures interact with the partially exposed receptor machinery responsible for the switch between the active and inactive states. Depending on how they have been designed to interact, one can predispose these substructures to favour either one state or the other; thus privileged structures can be used to create either agonists or antagonists. In terms of the mechanism of recognition, the region that the privileged structures bind to are rich in aromatic residues, which explains the prevalence of aromatic groups and atoms such as sulphur or halogens in many of the ligands. Finally, the approach described here can be used to design drugs for orphan receptors whose function has not yet been established experimentally.

1 Introduction

An appropriate way to begin this paper would be to quote the following official IUPAC definition (IUPAC 1999): "Privileged Structure: Substructural feature which confers desirable (often drug-like) properties on compounds containing that feature. Often consists of a semi-rigid scaffold which is able to present multiple hydrophobic residues without undergoing hydrophobic collapse, e.g. diazepam in which the diphenylmethane moiety prevents association of the aromatic rings."

This definition will be adhered to, although in the literature there are examples where the concept loses some of its intended generality or universality by being used to explain exclusive behaviour in the context of some rather narrow set of targets, e.g. aryl piperazines being selective for, in different cases, dopamine (Sukalovic et al. 2005) and melanocortin (Dyck et al. 2003) targets, whereas of course, this class of compounds are eminently suitable as ligand substructures in many different contexts (Mason et al. 1999). The main idea is intended to be that the part of the ligand containing the privileged structure will manifest some binding or activation properties against quite a wide range of targets, whereafter the drug designer then proceeds to engineer-in those extra features that are more specific for the target of her or his choice (Mason et al. 1999; DeSimone et al. 2004; Costantino and Barlocco 2006; Bakshi et al. 2006; Nieto et al. 2005; Fisher et al. 2005; Guo and Hobbs 2003; Nicolaou et al. 2000a, 2000b, 2000c). The privileged structure is thereby being used as a scaffold which has been shown to have good chemical properties and which may with advantage also be drug-like [however one may choose to define that (Frimurer et al. 2000)].

This paper will not dwell so much on the privileged structures themselves and the different varieties on offer, nor on how they were discovered, since these have been thoroughly reviewed elsewhere (Mason et al. 1999; DeSimone et al. 2004; Costantino and Barlocco 2006; Bakshi et al. 2006; Nieto et al. 2005; Fisher et al. 2005; Guo and Hobbs 2003; Nicolaou et al. 2000a, 2000b, 2000c). The focus here is rather on what the mechanistic principles are behind how they function: what general features of the targets are exploited by the general features possessed by these privileged substructures. In order to illustrate these mechanistic principles better, it is expedient to consider a family of targets whose members belong to a superfamily with the same overall fold, but which display considerable diversity. This diversity accounts for the manifold different biological functions that the members display, and it also represents both the challenge and the opportunity that the drug designer is faced with. The question is then, are there any general features alongside all this diversity such that we can utilize previous ligand fragments or substructures that have earlier been found practical to deal with? The family chosen here is that of the G-protein-coupled receptors (GPCRs).

Mention of a previously published study (Bondensgaard et al. 2004) may serve as the best way to conclude this introduction. In that work, the conserved features in a large subset of the GPCRs—Class A, or rhodopsin-like receptors—were examined and it was concluded that in parallel with the occurrence of conserved features in ligands containing privileged structures, there were conserved regions of the ligand-binding pocket that serve as a privileged subpocket. This hypothesis will be further examined here. In particular, an explanation is in order for how it comes about that there is indeed such a privileged subpocket suitable for binding the privileged substructures of the ligand, yet the receptor has not been exposed to the selection pressure of these synthetic ligands for any length of evolutionary history.

2 Construction of Models of Class A GPCRs

GPCRs and membrane proteins in general are grossly underrepresented in the database of protein 3D structures (XRC and/or NMR). The reasons for this are that membrane proteins typically are produced in very small amounts compared with, for example, enzymes or antibodies, and that, because their membrane environment is critical for their stability and survival, experimental methods to solubilize them and find the right formulation that is conducive to crystal formation are far more tricky than for water-soluble globular proteins. This disparity is particularly infelicitous when it is remembered that GPCRs are among the most important components of biological signal transduction processes in eucaryotes, making them the most important set of targets for pharmacological intervention in combating disease.

The only crystal structure of GPCRs are different crystalline variants of the same protein, bovine rhodopsin (Palczewski et al. 2000; Li et al. 2004; Ruprecht et al. 2004). This protein is a representative structure for Class A GPCRs; however it is not really suitable as a canonical structure of this class (Bywater 2005). Nevertheless, it is the only GPCR that has a crystal structure, and therefore the only choice available for use as a template for constructing models of other Class A GPCRs by homology modelling.

In this work, the WHAT IF program (Vriend 1990) has been used for homology modelling and YASARA (Krieger and Vriend 2002) for molecular graphics.

3 Detection of Conserved Regions in Class A GPCRs

Multiple alignments of over 1,600 Class A GPCRs are documented in the GPCRDB. From these alignment data, it is possible to compute a number of useful parameters related to residue type conservation, entropy[1] and correlated mutations and to produce phylogenetic trees[2]. It has been observed (Bondensgaard et al. 2004; Oliveira and Vriend 2003) that regions of lowest entropy (most conserved) are located in a region close to and within the core of the receptor in a subregion that extends downwards to the G-protein binding region as well as upwards so as to form a lining to the bottom of the classical biogenic amine-binding site (see next section). Residues at other sites have lower degrees of conservation, reflecting their different roles (e.g. binding of specific ligands). See Figs. 1, 2 and also Fig. 2 in Bondensgaard et al. 2004.

4 Ligand-Binding Regions in Class A GPCRs

For the purposes of this discussion, we are interested in residues that are in contact with ligands. This poses a problem in that the Class A receptors recognize ligands of many different types and size ranges, from small biogenic amines up to peptides and even proteins. The small ligands typically bind in the well-characterized ligand-binding pocket lodged between TMs 3, 5, 6 and 7 (Strader et al. 1987, 1989). In ad-

[1]Shannon entropy, or informational entropy, defined in the normal way as:

$$S_p = -\Sigma_{i=1,20} f_{p(i)} \ln\left[f_{p(i)}\right] \quad (1)$$

where $f_{p(i)}$ is the relative frequency of residue type i at position p.

[2]These are more properly called dendrograms since the coverage of evolutionary space is not comprehensive enough to warrant use of the term phylogenetic.

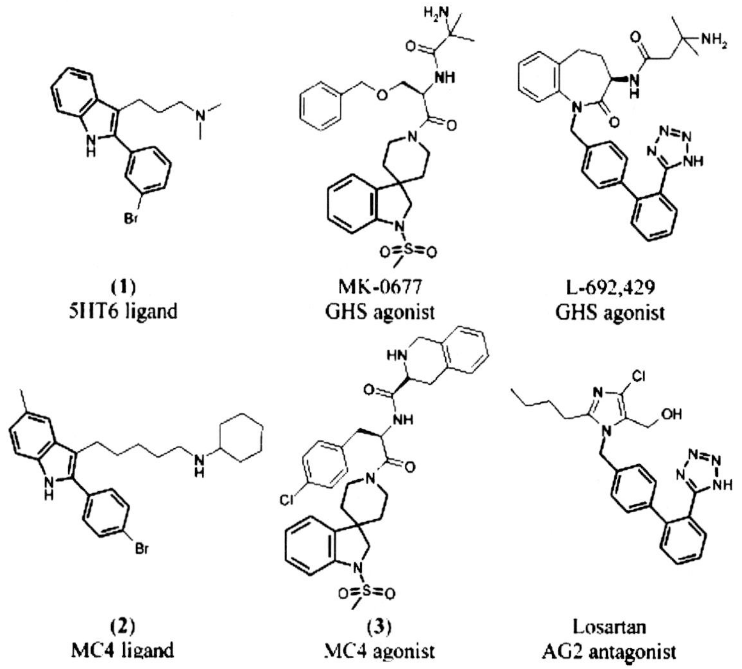

Fig. 1. Privileged structures mentioned in this study. (From Bondensgaard et al. 2004)

dition, sites near the top of TM4 can be involved in ligand binding (Bywater 2005; Suryanarayana 1992), as will also be encountered in this work. We define the small ligand-binding pocket as being made up of those residues which contact these small ligands. There is no conflict with the need to define a ligand-binding pocket for larger ligands such as peptides, since these too occupy the small ligand-binding pocket but of course, their greater size means that they extend well outside of this pocket, upwards and out into the external loop regions where many further contacts are made. Many of these contacts contribute additionally to both activity and specificity. Nevertheless, the peptide ligands invariably utilize the small ligand pocket as well (see fuller discussion with references in Bywater 2005). Furthermore, most synthetic drugs

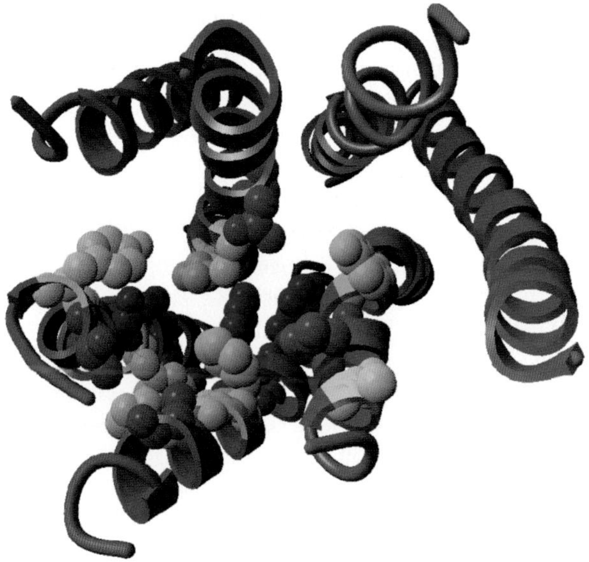

Fig. 2. Consensus structure of the aminergic receptors from the extracellular side colour-coded by entropy (*red* – highest entropy → *blue* – lowest entropy)

designed to mimic the function of these peptides are usually smaller, more compact, and they usually have the small ligand-binding pocket as their destination on docking. In this work, the residues that line this pocket are: 3.7, 3.8, 3.11, 3.12, 4.16, 4.20, 5.5, 5.6, 5.9, 5.10, 6.23, 6.27, 6.30, 7.4, 7.5, 7.8, 7.9, 7.11.

The binding pocket described above does not occupy all of the cleft between the TMs mentioned: it is possible to penetrate deeper into this cleft. Endogenous ligands do not normally do so, but many synthetic ligands do. The latter are in any case usually somewhat larger than the endogenous ligand and can, when superposed on the endogenous, extend both above and below it. In this work, the deeper recesses of the overall binding region reached by these ligands are treated as a separate subpocket, here given the attribute "privileged" in anticipation of the studies carried out into the nature of this subpocket. In a previous study (Bondensgaard et al. 2004), a number of ligands (e.g., those shown in

Fig. 1), were docked into their corresponding receptors. This enabled the residues that form the deeper cleft to be identified as: 3.12, 3.15, 3.16, 3.19, 4.19, 4.23, 5.5, 5.6, 5.9, 5.10, 5.13, 5.14, 6.16, 6.20, 6.23, 6.24, 6.27. Note that there is a certain amount of overlap between the two sets. The justification for the choice of residue positions for the privileged set comes from published (Bondensgaard et al. 2004) docking studies in which the ligands of Fig. 1 were docked into various Class A GPCRs. The contacts that arise from these dockings are summarized in Table 1, and it can immediately be seen that the ligands are utilizing the above-mentioned set of privileged residues. The receptor residues involved are predominantly aromatic and it is therefore hardly surprising that privileged structures almost invariably contain an aromatic moiety. It is well known than aromatic–aromatic interactions are important in maintaining protein structure and in attracting aromatic ligands (Burley and Petsko 1985; Samanta et al. 1999, 2000; Singh and Thornton 1985; Thomas et al. 2002), and sulfur (Pal and Chakrabarti 2001), halogens (Adams et al. 2004) and cations (Mo et al. 2002) readily form complexes with aromatic groups.

Table 1 Essential ligand-receptor contacts for the six ligands in Fig. 1 (data from Bondensgaard et al. 2004)

Receptor	Ligand	Ligand ↔ receptor contacts
5HT6R	1	Indole ↔ F5.14/F6.16/W6.20 Br ↔ F5.5
		Aminoalkyl ↔ D3.11
MC4R	2	Indole ↔ F5.14/F6.16/W6.20 Br ↔ F4.23
		Aminoalkyl ↔ D3.8
GHSR	MK-0677	Spiro ↔ F5.14/F6.16/F6.20/F6.23
		NSO_2CH_3 ↔ H6.24 Aminoalkyl ↔ E3.12
MC4R	3	Spiro ↔ F5.14/F6.16/F6.20/F6.23
		NSO_2CH_3 ↔ H6.24 Aminoalkyl ↔ E3.12
AG2R	Losartan	Biphenyl ↔ V3.11/ S3.12/Tetrazole ↔ K5.9/H6.23
GHSR	L-692.429	Biphenyl ↔ F5.14/F6.16/W6.20
		Tetrazole ↔ K5.9/N5.10/Q6.24

↔ Means there is a contact between a ligand moiety and the specifiedresidue(s), numbered with local TM position numbering convention as elsewhere in this chapter

5 Partitioning Sequences into Functionally Defined Segments

In order to compare the two sets of ligand binding residues over several members of Class A at once, an extraction procedure (Bywater 2005; Jacoby 2002) is used whereby the multiple sequences are partitioned according to function (see Fig. 3). The segments of the multiple alignment created in this way are then concatenated to produce a new, shorter, multiple alignment. While the sequences in this alignment are no longer contiguous in the sequence as dictated by the genotype, they encapsulate phenotypic information (a protein has several functions or phenotypes, and ligand recognition is only one of them). Thus a phylogenetic analysis, or dendrogram, of different segments of the sequence will reveal different clustering of the receptor types, compared to each other (see Fig. 3) and of course compared with the entire global sequence (not shown).

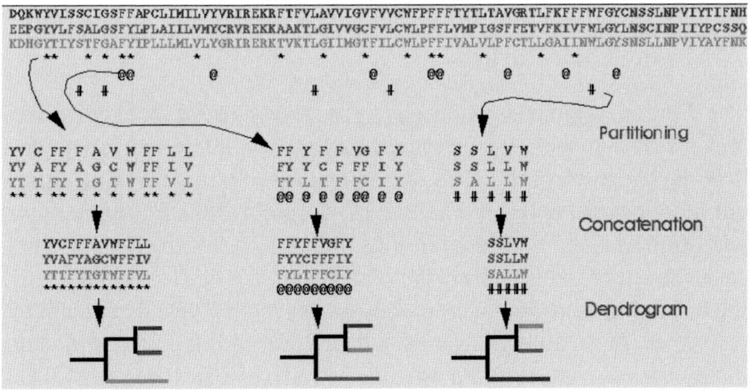

Fig. 3. Any protein must fulfil many different tasks. Since each of these evolves under different selection pressures, they can be subjected to phylogenetic analysis by partitioning the full-length sequence into sequence subsets responsible for each function

6 Nature of the Two Binding Subregions

It is immediately apparent from Table 2 that there are major differences between the two subpockets in respect of both conservation and chemical character. The specificity set is, as expected, much more diverse in sequence (Table 2, column 2), while the privileged set is both more conserved and hydrophobic, and above all, aromatic in character [Table 2, column 3. Green indicates hydrophobic residues. This is only provided visually here, but of course it can easily be quantified (this can be left as an exercise for the reader)]. The nature and the role of the aromatic residues has been discussed before (Bondensgaard et al. 2004) and these residues correspond to those that act as the core machinery in the transition between active and inactive states of the receptor (Gouldson et al. 2004). In both states, part of this hinge is exposed to the binding pocket, but of course, the exposed regions/residues will be different in the active and inactive states. This both explains why privileged structures turn up in antagonists as well as agonists and provides an opportunity for some selective engineering to steer the activity one way or the other by appropriate design of the privileged part of the ligand.

7 Clustering and Potential Cross-Reactivity of Ligands and Their Targets

The different segments, produced (e.g. as shown in Fig. 2) have evolved under different selection pressures, and we are interested in knowing how the members of the class or family cluster in respect of the different functions/phenotypes. The two dendrograms for the specificity set and privileged set are shown in Figs. 4 and 5, respectively. In the first case, the clustering is entirely familiar, the opiates cluster together as do the histamines and the other families, as expected. The clustering in the privileged set is entirely different. It even looks as if it could be random, but by the very nature of the way in which the residues are selected and then subjected to dendrographic analysis, it cannot be. What it reveals is that there is a clustering which can be exploited, if so desired, to take a privileged structure that has been successful in the context of one target and reuse it with another. Note that in this case the clustering, while

Table 2 Concatenated residues after partitioning into column 2, ligand binding pocket, and column 3, "privileged subpocket"

Receptor	Ligand binding pocket	"Privileged sub-pocket"
BLTR	CHCGSAHLEAYNERNIA	GMYVLVHLEAGFFWYHN
CLT1	STLYVSLHSLYRHVVLS	YLYITFLHSLGFFFYHR
EDG1	REMFSGHYFCLLDAELV	FLSVLPHYFCVFFWLFL
EDG2	RQIDAGSYFWGLDEKLL	DLTVMPSYFWFNFWGLL
P2Y1	QRFHVIYSTTFKNYQRG	HLYIAIYSTTMFFYFHK
P2Y2	CFKYVCMIAEYFYHPLC	YMYIAFMIAEGFFFYHF
P2Y6	VRFYVCYGLTFKYYKRP	YLHIQTYGLTGFFFFHK
P2Y9	SGFLVGITIEYLYYPLC	LIYMGAITIEGFFFYNL
PD2R	FAMSGFGYLYVAGDLLR	SGLTKYGYLYLMFSVIA
CB1R	KLVTAADELMLMDFASM	TFTVIPDELMIGLWLLM
SPR1	CGLYESINRFYLRYHLL	YIYVTFINRFGFFFYHL
TAR1	HTDIPASGTFFTDNDIW	ISSIFMSGTFSFFWFFT
5H1A	FIDVGSYTSTFALGANW	VCTIIPYTSTAFFWFFA
5H2A	WIDVSSFVGSFNALNVW	VSTIIIFVGSSFFWFFN
5H4	RTDVPSYACSFNDWTLW	VTTIIPYACSAFFWFFN
5H5A	WIDVSSYASTFESKSLW	VCTIIPYASTAFFWFFE
5H6	WTDVASFVASFNQFDTW	VCSIAPFVASTFFWFFN
5H7	FIDVSTYTSTFSRERLW	VCTIIPYTSTAFFWFFS
A1AA	WADVSSYVSAFMGFKFW	VCTIIPYVSASFFWFFM
A2AA	YLDVSSYVSCFYTFKFW	VCTIIPYVSCSFFWFFY
B1AR	WTDVSSYASSFNKFVNW	VVTIVPYASSSFFWFFN
DADR	WVDISSYASSFNLFDVW	ISTIIPYASSSFFWFFN
D2DR	FVDVSSFVSSFHNYSTW	VCTIILFVSSSFFWFFH
ACM1	WLDYSWITTAYVSWEYW	YSNVLAITTAAFFWYNV
HH1R	WLDYSVFKTAYFIHMIW	YSTIWIFKTANFFWYFF
HH2R	YTDVSSYGDGYFREALW	VCTILSYGDGTFFWYFF
HH3R	WLDYAYFLASYMRYEFW	YCTALAFLASEFFWYTM
HH4R	WLDYANILTSYTLYRFW	YCTVVMILTSEFFWYST
PAFR	AGFFIAIIFIHQWHQLC	FTYVAFIIFISFFFHHQ
NY1R	NPQCASYTLLLNFFLHL	CITISFYTLLQYFWLTN
NTR1	YYRDSTVINTYRFYMNA	DTYALMVINTSFFWYHR
GHSR	FQSEAAVMVSFRFNLFV	ETYVSIVMVSFFFWFHR
NK1R	HNPIAAYHVTFFPYLMW	IVFILQYHVTIYFWFHF
OPRD	VLDYAGTKVFIVWLHIA	YMFIVITKVFAFFWIHV
OPRK	VIDYSGMKVFIIEYYIA	YMFIVAMKVFAFFWIHI
OPRM	VIDYSGLKVFIVKWHIA	YMFIIVLKVFAFFWIHV
OPRX	VIDYAGFAIFVVQLRTA	YMFTVVFAIFSFFWVQV
MC4R	IDICCGAVCLFLYFNLI	CLLISFAVCLMFFWFFL
AA1R	VAVLSGMVNFLNTTYIF	LTQIVPMVNFWVFWLHN
MCHR1	ITDASIFTQFYQQYNIS	AQFTSVFTQFAFFWYYQ
OPSD	EGATAAFVMFYAIMTAF	TGELCPFVMFHFFWYAA
CAMN	WLDVSSYASSFNRFDFW	VCTIIPYASSSFFWFFN
CPEP	VTYFASQTIFYLNMQET	FFFILGQTIFGFFWYNL
CONSENUS	D SS S	V GFFWYNL

Hydrophobic residues are in green. Note how the residues in the privileged set are both more hydrophobic and more conserved. *CAMN* means residues taken from the global consensus for the aminergic receptors, *CPEP* the corresponding set of residues for the Class A peptide receptors. *Consensus* is the local consensus for these two partitioned sets. For the ligand-binding pocket set there is very little or no conservation, but the positions of the canonical D residue and the SSxxS motif (both typical of the aminergics) are shown. Conserved residues for the privileged set are shown in *Consensus*, column 3. The receptors in this list are a truncated (for space reasons) of the collection that are shown in Figs. 4 and 5

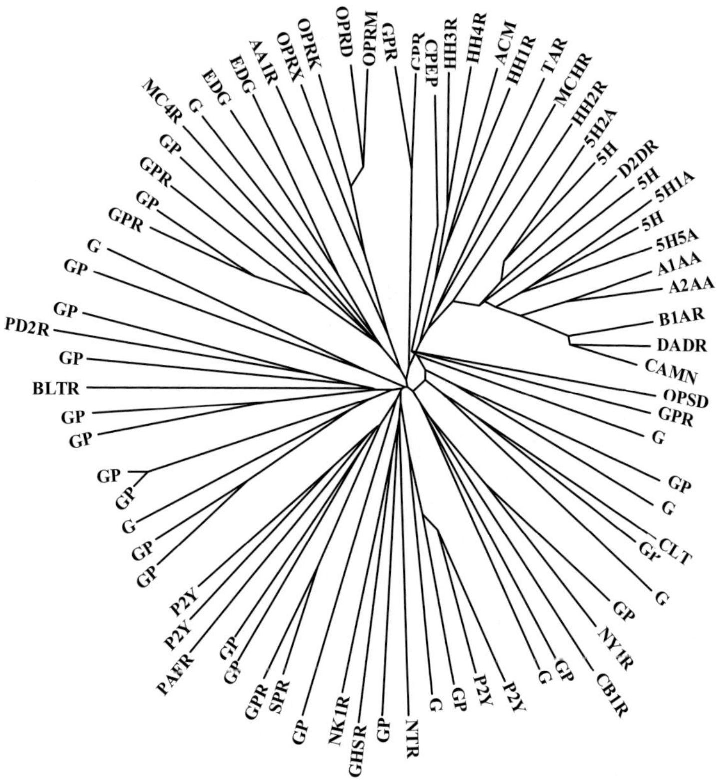

Fig. 4. Unrooted phylogenetic tree for GPCRA sequences partitioned according to ligand specificity. (Treecon program, Van de Peer and De Wachter 1997)

certainly not entirely random, is circumstantial, since the receptors have not evolved under the selection pressure of this privileged structure.

A suitable drug-design strategy might therefore be to use the specificity clustering to design the specific part of the ligand, which is then grafted onto a privileged structure chosen with the aid of information from the privileged clustering.

When these clusters are inspected, it is easy to identify receptor types that respond to the same ligand. This can give valuable clues as to which

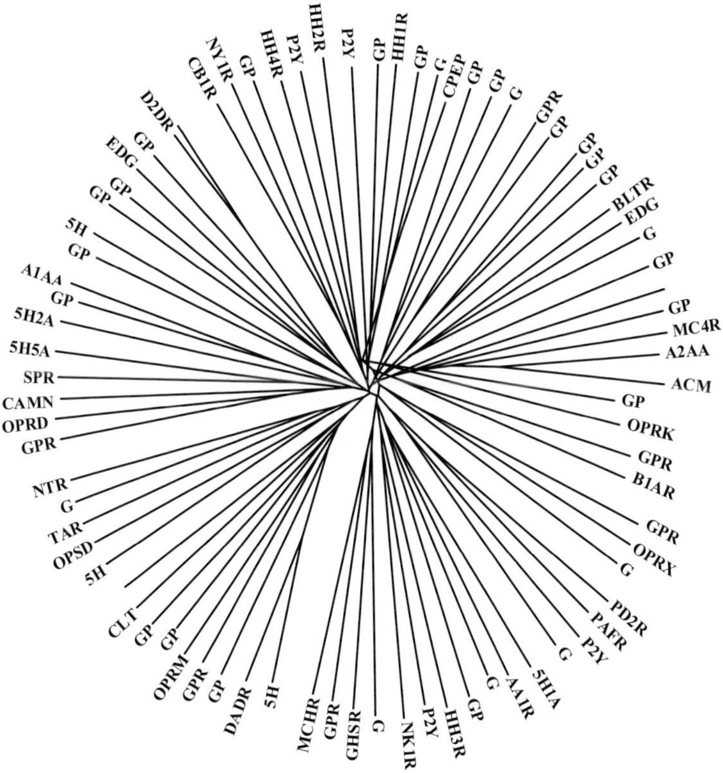

Fig. 5. Unrooted phylogenetic tree for GPCRA sequences partitioned according to privileged structure binding propensity. (Treecon program, Van de Peer and De Wachter 1997)

receptor types may also inadvertently respond to the same synthetic ligands (maybe a different residue subset, reflecting the docking of the synthetic ligand, rather than the endogenous ligand). Conversely, one may want to exploit this kind of information in order to design drugs that are active at more than one target, so-called dirty drugs, but a term like "multipotent" would be preferable.

Another useful feature of this kind of focused phylogenetic clustering is that orphan receptors that cluster close to receptor types with known function can be suspected of having a propensity to bind the same ligand. This would not be noticed if full-length sequence alignments had been used. This is in a sense an *in silico* type of deorphanization.

8 Conclusions

An explanation has been provided as to why certain types of substructure, privileged structures, recur frequently in pharmacologically successful synthetic ligands. One feature which favours their use is the evident drug likeness of these privileged structures. The other is that they represent a generic substructure that matches commonly recurring conserved structural motifs, privileged subpockets, in the target proteins, which may otherwise be quite diverse in sequence and function. By identifying the different residues that line the different regions of the ligand binding pocket, that where the privileged structure docks and that where the substructure containing the specificity message resides, it is possible, for example by appropriately partitioning multiple sequence alignments and examining the selected regions phylogenetically, to identify which other receptors may respond to the privileged substructure or the specificity substructure, respectively. Undesired side effects (or even desired ones: the multipotent drug concept) can be anticipated in this way. Privileged structures interact with the partially exposed machinery involved in the switch between the active and inactive states. Depending on how the privileged ligands have been designed to interact, one can predispose these substructures to favour either one state or the other; thus privileged structures can be used to create either agonists or antagonists. The residue side-chains involved in the activation machinery are predominantly aromatic, which explains the prevalence of aromatic moieties and of halogen or sulphur atoms in the privileged structures. Finally, the approach described here lends itself to the design of drugs for orphan receptors whose function has not yet been established experimentally.

References

Bakshi RK, Hong Q, Tang R, Kalyani RN, Macneil T, Weinberg DH, Van der Ploeg LH, Patchett AA, Nargund RP (2006) Optimization of a privileged structure leading to potent and selective human melanocortin subtype–4 receptor ligands. Bioorg Med Chem Lett 16:1130–1133

Bondensgaard K, Ankersen M, Thogersen H, Hansen BS, Wulff BS, Bywater RP (2004) Recognition of privileged structures by G-protein coupled receptors. J Med Chem 47:888–899

Burley SK, Petsko GA (1985) Aromatic-aromatic interaction: a mechanism of protein structure stabilisation. Science 229:23–28

Bywater RP (2005) Location and nature of the residues important for ligand recognition in Class A G-Protein coupled receptors. J Mol Recogn 18: 60–72

Costantino L, Barlocco D (2006) Privileged structures as leads in medicinal chemistry. Curr Med Chem 13:65–85

DeSimone RW, Currie KS, Mitchell SA, Darrow JW, Pippin DA (2004) Privileged structures: applications in drug discovery. Comb Chem High Throughput Screen 7:473–494

Dyck B, Parker J, Phillips T, Carter L, Murphy B, Summers R, Hermann J, Baker T, Cismowski M, Saunders J, Goodfellow V (2003) Aryl piperazine melanocortin MC4 receptor agonists. Bioorg Med Chem Lett 13:3793–3796

Fisher MJ, Backer RT, Husain S, Hsiung HM, Mullaney JT, O'Brian TP, Ornstein PL, Rothhaar RR, Zgombick JM, Briner K (2005) Privileged structure-based ligands for melanocortin receptors-tetrahydroquinolines, indoles, and aminotetralines. Bioorg Med Chem Lett 15:4459–4462

Frimurer TM, Bywater RP, Naerum L, Nørskov-Lauritsen L, Brunak S (2000) Discriminating "drug-like" from "non drug-like" molecules: improving the odds. J Chem Inf Comput Sci 40:1315–1324

Gouldson PR, Kidley N, Bywater RP, Psaroudakis G, Brooks HD, Diaz C, Shire D, Reynolds CA (2004) Towards the active conformations of rhodopsin and the β-2-adrenergic receptors. Proteins Struct Funct Genet Bioinformat 56:67–84

Guo T, Hobbs DW (2003) Privileged structure-based combinatorial libraries targeting G protein-coupled receptors. Assay Drug Dev Technol 1:579–592

IUPAC (1999) Glossary of terms used in combinatorial chemistry. Pure Appl Chem 71:2349–2365

Jacoby E (2002) A novel chemogenomics knowledge-based ligand design strategy–application to G-protein coupled receptors. Quant Struct Activity Relat 20:115–122

Krieger E, Vriend G (2002) Models@Home: distributed computing in bioinformatics using a screensaver-based approach. Bioinformatics 18:315–318
Li J, Edwards PC, Burghammer M, Villa C, Schertler GF (2004) Structure of bovine rhodopsin in a trigonal crystal form. J Mol Biol 343:1409–1438
Mason JS, Morize I, Menard PR, Cheney DL, Hulme C, Labaudiniere RF (1999) New 4-point pharmacophore method for molecular similarity and diversity applications: overview of the method and applications, including a novel approach to the design of combinatorial libraries containing privileged substructures. J Med Chem 42:3251–3264
Mo Y, Subramanian G, Gao J, Ferguson DM (2002) Cation-π interactions: an energy decomposition analysis and its implications in δ-opioid receptor-ligand binding. J Am Chem Soc 124:4832–4837
Nicolaou KC, Pfefferkorn JA, Barluenga S, Mitchell HJ, Roecker AJ, Cao GQ (2000a) Natural product-like combinatorial libraries based on privileged structures. The "libraries from libraries" principle for diversity enhancement of benzopyran libraries. J Am Chem Soc 122:9968–9976
Nicolaou KC, Pfefferkorn JA, Mitchell HJ, Roecker AJ, Barluenga S, Cao GQ, Affleck RL, Lillig JE (2000b) Natural product-like combinatorial libraries based on privileged structures. Construction of a 10000-membered benzopyran library by directed split-and-pool chemistry using nanokans and optical encoding. J Am Chem Soc 122:9954–9967
Nicolaou KC, Pfefferkorn JA, Roecker AJ, Cao GQ, Barluenga S, Mitchell HJ (2000c) Natural product-like combinatorial libraries based on privileged structures. General principles and solid-phase synthesis of benzopyrans. J Am Chem Soc 122:9939–9953
Nieto MJ, Philip AE, Poupaert JH, McCurdy CR (2005) Solution-phase parallel synthesis of spirohydantoins. J Comb Chem 7:258–263
Oliveira L, Paiva PB, Paiva AC, Vriend G (2003) Sequence analysis reveals how G protein-coupled receptors transduce the signal to the G protein. Proteins 52:553–560
Pal D, Chakrabarti P (2001) Non-hydrogen bond interactions involving the methionine sulfur atom. J Biomolec Struct Dynamics 19:115–128
Palczewski K, Kumasaka T, Hori T, Behnke CA, Motoshima H, Fox BA, Le Trong I, Teller DC, Okada T, Stenkamp RE, Yamamoto M, Miyano M (2000) Crystal structure of rhodopsin: a G protein-coupled receptor. Science 289:739–745
Ruprecht JJ, Mielke T, Vogel R, Villa C, Schertler GFX (2004) Electron crystallography reveals the structure of metarhodopsin I. EMBO J 23:3609–3620
Samanta U, Pal D, Chakrabarti P (1999) Packing of aromatic rings against tryptophan residues in proteins. Acta Crystallographica D55:1421–1427

Samanta U, Pal D, Chakrabarti P (2000) Environment of tryptophan side chains in proteins. Proteins 38:288–300
Singh J, Thornton JM (1985) The interactions between phenylalanine rings in proteins. FEBS Lett 191:1–6
Strader CD, Sigal IS, Register RB, Candelore MR, Rands E, Dixon RA (1987) Identification of residues required for ligand binding to the beta-adrenergic receptor. Proc Natl Acad Sci U S A 84:4384–4388
Strader CD, Candelore MR, Hill WS, Sigal IS, Dixon RA (1989) Identification of two serine residues involved in agonist activation of the beta-adrenergic receptor. J Biol Chem 264:13572–13578
Sukalovic V, Zlatovic M, Andric D, Roglic G, Kostic-Rajacic S, Soskic V (2005) Interaction of arylpiperazines with the dopamine receptor D2 binding site. Arzneimittelforschung 55:145–52
Suryanarayana S, von Zastrow M, Kobilka BK (1992) Identification of intramolecular interactions in adrenergic receptors. J Biol Chem 267:21991–21994
Thomas A, Meurisse R, Charloteaux B, Brasseur R (2002) Aromtaic side-chain interactions in proteins. Proteins 48:628–634
Van de Peer Y, De Wachter R (1997) Construction of evolutionary distance trees with TREECON for Windows: accounting for variation in nucleotide substitution rate among sites. Comput Appl Biosci 13:227–230
Vriend G (1990) WHAT IF: a molecular modelling and drug design program. J Mol Graph 8:52–56

Ernst Schering Foundation Symposium Proceedings, Vol. 2, pp. 93–103
DOI 10.1007/2789_2006_005

Published Online: 16 May 2007

Designing Compound Libraries Targeting GPCRs

E. Jacoby(✉)

Novartis Institutes for BioMedical Research, Discovery Tehnologies, Lichtstrasse 35, 4056 Basel, Switzerland
email: *edgar.jacoby@novartis.com*

Abstract. The design of compound libraries targeting GPCRs is of primary interest in pharmaceutical research because of their important role as signaling receptors and the herewith linked dominant place in the discovery portfolios. In the present symposium chapter, we outline GPCR compound library design strategies recently followed by our group and discuss them in a more general context.

1 Compound and Design Strategies Targeting GPCRs

As GPCRs are extremely versatile receptors for extracellular messengers as diverse as biogenic amines, purines and nucleic acid derivatives, lipids, peptides and proteins, odorants, pheromones, tastants, ions such as calcium and protons, and even photons in the case of rhodopsin, the design of compound libraries targeting GPCRs needs to address this diversity of molecular recognition (Jacoby et al. 2006).

Generally, the design of deorphanization libraries can be distinguished from targeted lead-finding libraries.

Given the broad chemical diversity of the hormones that are recognized by GPCRs, deorphanization libraries try to cover as many known active chemical classes as possible. The term "surrogate agonist library" is also appropriate given that the purpose of these libraries is to find a chemical compound that selectively activates a given orphan receptor of interest (Wise et al. 2004). Typically, compounds identical or similar to previously identified GPCR agonists are included with approved drugs and other reference compounds with known bioactivity, such as primary metabolites (e.g., KEGG compound set) or commercially available compilations, such as the Tocris LOPAC, the Prestwick, or the Sial Biomol sets. In addition to HPLC fractionations of tissue extracts to identify new peptides and metabolites, protein mimetic libraries are of interest, including β-turn/α-helix mimetics together with random or designed peptide libraries based on the bioinformatics analysis of putatively secreted peptides and protein hormones defined in the genome. Typically, the size of deorphanization or surrogate sets is on the order of a few thousand well-characterized compounds amenable for medium-throughput screening.

The design of lead-finding libraries follows the same molecular mimicry principles and makes the best use of the substantial medicinal chemistry knowledge generated during the last few decades around GPCR compounds together with more modern concepts, including lead/drug likeness and computational combinatorial library design (Klabunde and Hessler 2002; Bleicher et al. 2004; Crossely 2004; Jimonet and Jäger 2004).

Although focused GPCR library design concepts target, in general, the classical binding sites, design concepts of bivalent ligands and allosteric ligands are expected to become more important in the future given the anticipated progress in the understanding of the GPCR oligomerization phenomenon (Halazy 1999). The experience with focused libraries and screening sets for GPCRs is very positive and hit rates of up to 1%–10% can be expected with library sizes of 500–2500 compounds, when the libraries are designed toward new members with expected conserved molecular recognition. Peptide and protein mimetics libraries including β-turn/α-helix mimetics are recognized to be of central importance (Hruby 2002; Tyndall et al. 2005). A number of important hormones, such as angiotensin, bradykinin, CCK (cholecystokinin), MSF (melanin-stimulating factor) and SST (somatostatin) make their key recognition via specific β-turn motifs. Others, such as CRF (corticotrophin-releasing factor), PTH/PTHrP, NPY (neuropeptide Y), VIP (vasoactive intestinal peptide), or GHRF (growth hormone-releasing factor) interact via α-helix motifs (Webb 2004).

The use of privileged substructures or molecular master keys, whether targeting class-specific or mimicking protein secondary structure elements, is an accepted concept in medicinal chemistry. The privileged structure approach emphasizes molecular scaffolds or selected substructures that are able to provide high-affinity ligands (agonist or antagonists) for diverse receptors and originates from work at Merck Research Laboratories on the design of benzodiazepine-based CCK antagonist, where the previously known κ-opioid Tifluadom was identified as a lead structure (Evans et al. 1988). A number of recent literature reviews provide impressive reference repertoires of empirically derived privileged structures, most notably the spiropiperidines, biphenyltetrazoles, benzimidazoles, and benzofurans (Guo and Hobbs 2003).

The development of cheminformatics methods and procedures enabling the automatic identification and extraction of privileged structures is especially needed in the context of generating knowledge from HTS data. Based on the molecular framework approach developed by Bemis and Murcko (1996), we recently initiated a systematic analysis using a reference compound and target information. Using the framework analysis as implemented in the Scitegic Pipeline Pilot software, we designed a data pipelining protocol that generates frequency analyses based on the input of the various reference sets. The approach is illustrated in Fig. 1 for monoamine GPCR ligands. The listed scaffolds are of immediate interest for targeted library design approaches.

A different type of fragment-based design method was developed previously by our group when designing compound libraries targeting monoamine-related GPCRs. Based on the central chemogenomics principle that similar ligands bind to similar targets and that ligands of close homologous receptors are generally considered as putative starting points in lead-finding programs for receptors for which no specific ligands are yet known, we proposed a chemogenomics knowledge-based combinatorial library design strategy for lead finding (Jacoby 1999, 2001). The strategy is founded on the integration of both the de-

Fig. 1. Analysis of a privileged scaffold-target matrix of monoamine GPCR ligands. For each GPCR ligand assigned in the MDDR (MDL Drug Data Report) database to a specific monoamine GPCR subtype, the Bemis-Murcko framework was generated. The lists of frameworks were then combined and duplicates were eliminated. The comprehensive list of unique frameworks defines the row vector of the matrix, and the GPCR subtypes were arranged to the column vector. The matrix elements were assigned according to the number of compounds reported including a given framework for a given subtype. In addition, for each framework the total number of monoamine GPCR subtypes addressed was added and summarized in the frequency column; the rows were then sorted by decreasing frequency. The structures of the seven most represented frameworks together with the addressed monoamine GPCR subtypes are shown

convolution of known modular ligands of homologous receptors into their component fragments and the structural bioinformatics comparison of the binding sites for the individual ligand fragments. In essence, in the ligand space, by the analysis of both the ligand architectures and the structures of the component one-site filling fragments of known ligands, it should be possible, by referring to the locally, most directly related and characterized receptors, to identify those component ligand fragments, which based on the binding site similarities are potentially best suited for the design of ligands tailored to the new target receptor. The strategy was presented in the context of designing the TAM (tertiary amine) combinatorial library directed toward monoamine-related GPCRs (see Fig. 2) for which the conserved aspartate residue $D_{3.32}$ in TM3 was demonstrated by two-dimensional mutation experiments to be responsible for the recognition of the charged amino group of monoamine ligands by their GPCRs.

Focusing on the central importance of the $D_{3.32}$ residue and using the $D_{3.32}X_{16}$(DE)R(YFH) motif in TM3 as a sequence signature defining relatedness to the monoamine GPCR subfamily, we identified, by database searches, 50 human GPCRs, which included seven orphan GPCRs (two of which are now known to correspond to pseudogenes) and which constituted the library's target repertoire originally sought. Later it was recognized that trace amine receptors—which conserve the $D_{3.32}$ residue as well as chemokine receptors, which lack the $D_{3.32}$

	5HT1A	5HT1B	5HT1D	5HT1F	5HT2A	5HT2B	5HT2C	5HT4	A1A	A2A	B1	B3	D1	D2	D3	D4	M1	M3	H2	Frequency
PS1	0	0	0	0	2	0	0	0	0	4	0	0	1	11	0	0	0	0	0	4
PS2	0	0	0	0	0	1	1	0	0	0	0	0	0	2	0	14	0	0	0	4
PS3	6	0	0	0	3	3	3	0	0	0	0	0	0	0	0	0	0	0	0	4
PS4	2	0	0	0	1	1	1	0	0	0	0	0	0	0	0	0	0	0	0	4
PS5	2	0	0	0	0	0	0	0	1	0	0	0	0	2	1	0	0	0	0	4
PS6	0	0	0	0	0	0	0	0	1	1	1	0	0	0	0	3	0	0	0	4
PS7	2	0	0	0	0	0	0	0	0	0	0	0	1	1	0	1	0	0	0	4

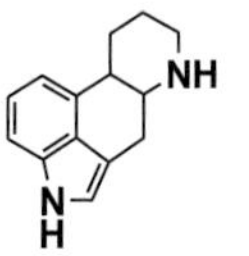

PS1

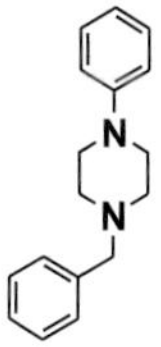

PS2

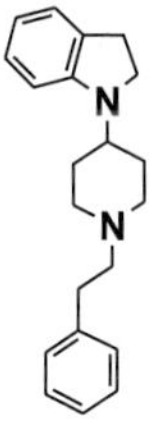

PS3

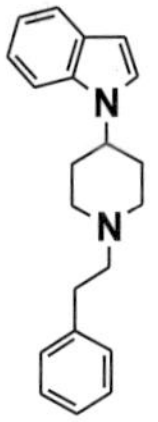

PS4

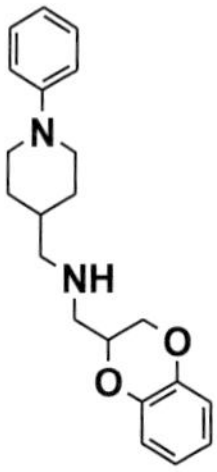

PS5

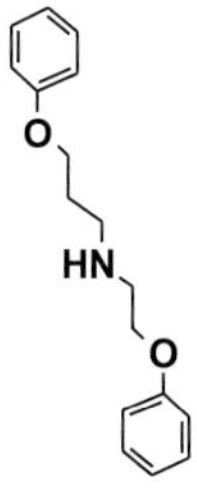

PS6

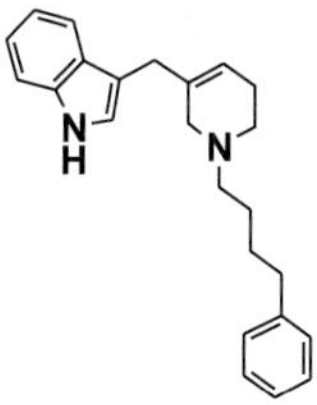

PS7

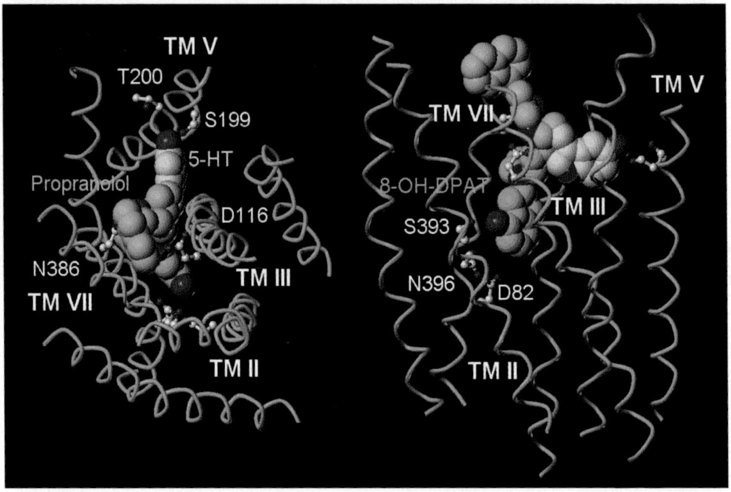

Fig. 2. Three ligand-binding sites model for monoamine-related GPCRs illustrated by a rhodopsin-based 3D model of the 5-HT_{1A} receptor (*left*, extracellular view; *right* side view). We recently proposed a three-binding-site hypothesis for the molecular recognition of ligands at monoamine GPCRs by combining: (*1*) analyses of the architectures of known monoamine GPCR ligands (see Fig. 3); (*2*) analyses of molecular models of the ligand-receptor interactions; and (*3*) structural bioinformatics analyses of the sequence similarities of the three distinct binding regions of one-site-filling ligand fragments within the monoamine GPCR family. For the 5-HT_{1A} receptor, which provided a template for the discussion of other related ligand–GPCR interactions, mutagenesis studies map three spatially distinct binding regions, which correspond to the binding sites of the small, one-site-filling ligands 5-HT (serotonin, *yellow*), propranolol (*cyan*), and 8-OH-DPAT (8-hydroyxy-*N,N*-dipropylaminotetralin, *green*). All three binding sites are located within the highly conserved 7TM domain of the receptor and overlap at the residue $Asp_{3.32}$ (D116) in TM3, which is the key anchor site for basic monoamine ligands. The three distinct binding sites are also reflected by the architectures of known high-affinity ligands, which crosslink two or three one-site filling fragments around a basic amino group

residue but where a corresponding glutamate residue $E_{7.39}$ in TM7 is responsible for the recognition of the tertiary amine chemotype—have to be considered based on molecular recognition principles as monoamine-like GPCRs. This extends the target repertoire to around 80 GPCRs, a significant part of all the class A GPCRs.

Databases of site-specific ligand fragments, which should be recombined on an appropriate scaffold to yield ligands, are the keystones of such a knowledge-based system. Their generation is theoretically possible through the deconvolution of the known ligands guided by SAR and by molecular similarity consideration. Given the promiscuity of some fragments (e.g., symmetric ligands), one should be cautious before drawing definitive conclusions about the actual positioning of the fragments. Pragmatically, these limitations to the generation of site-specific ligand fragment databases were approached by pooling fragments into multiple pools and by designing generic combinatorial libraries of known privileged active fragments around appropriate scaffolds. Prototype compounds obtained using this approach are shown in Fig. 3.

The TAM library was screened in a number of GPCR campaigns and high hit rates were especially observed for the monoamine and chemokine GPCRs. The TAM library includes many new combinations of known active fragments and privileged GPCR motifs. In addition to addressing new receptors, this should allow the discovery of interesting multireceptor profiles of potential pharmacological interest. The search of the antagonist for the 5-HT_7 GPCR, which has the 5-HT_{1A} receptors as neighbors in the sequence dendrogram, illustrates the successful use of the TAM library. Searching with $5HT_{1A}$ reference compounds using the Similog method (see below in this section) within the TAM library, we were able to identify a 10% hit rate ($pK_B < 5\ \mu M$) when only a biological assay with limited throughput capacity was available. The hits were arylpiperazines, which in follow-up studies were also active on other monoamine GPCRs.

This method is comparable to a fragment-based design method called thematic analysis developed by researchers at Biofocus for the design of focused GPCR libraries (Crossley 2004). In the Biofocus approach, SARs were analyzed in detail across the entire class A and class B GPCR family, and family-activity relationships were used to develop a new classification process based on the pairing of sequence themes and ligand structural motifs. A sequence theme is a consensus collection of amino acids within the central binding cavity and a motif is a specific

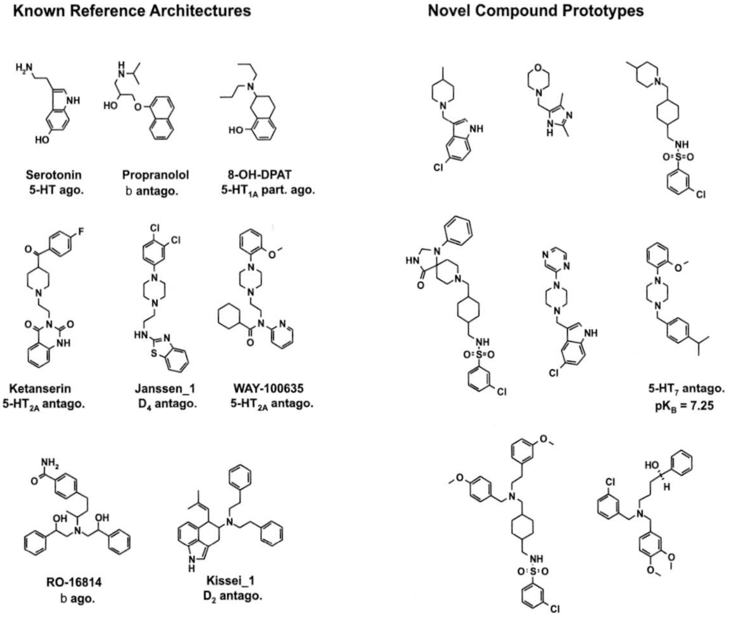

Fig. 3. Prototype structures of the Novartis TAM combinatorial libraries generated through reductive amination of selected aldehydes and secondary amines. The new structures for which examples are shown in the *right panel* were designed to be similar to known monoamine GPCR ligands for which examples are shown in the *left panel*. Ligands, which are the same size as the endogenous ligands, are called herein simple-one-site filling ligands. In addition to this natural architecture, ligands exist where two or three such simple ligand fragments are linked around a basic positively charged group: these ligands are called, correspondingly, double and triple ligands. All three architectures—simple, double, and triple—of known monoamine GPCR ligands are represented in the TAM library

structural element binding to such a particular microenvironment of the binding site. The analysis resulted in a compilation of themes and motifs, which today are used at Biofocus to generate focused discovery libraries and to increase the lead optimization efficiency for these targets. The individual compound libraries are targeting subsets of GPCRs, including orphans, that share a predefined combination of themes consisting of a central dominant theme and peripheral ancillary themes. The library scaffold is designed such that it complements the central theme and is amenable to incorporating a variety of structural motifs addressing the individual sequence themes. Each library, consisting of approximately 1000 compounds, can thus be thought of as representing a number of predefined themes, which are either present or absent in any given receptor.

Compared to the fragment-based approaches, several groups have developed knowledge-based library design strategies which are, in principle, based on Sir James Black's frequently quoted statement: "the most fruitful basis for the discovery of a new drug is to start with an old drug". The associated SOSA (selective optimization of side activities) approach is an additional, very successful medicinal chemistry concept where the atypical neuroleptics acting on a couple of GPCRs simultaneously provide a relevant illustration of the rationale (Wermuth 2004). The related CADD (computer-assisted drug design) methods make use of selected reference compound sets and molecular descriptors together with advanced cheminformatics methods to compare and rank the similarity of designed candidate molecules. As such, homology-based similarity searching was developed at Novartis as a cheminformatics similarity searching method able to identify not only ligands binding to the same target as the reference ligand(s), but also potential ligands of other homologous targets for which no ligands are yet known (Schuffenhauer et al. 2003). The method is based on the Similog descriptor, which describes molecules as counts of pharmacophore triplets formed by the individual non-hydrogen atoms and uses a centroid of the reference compounds to describe the distance to the candidate molecule. In a retrospective analysis, the method was shown to be highly effective for monoamine GPCRs and became an essential tool for the compilation of focused screening sets. Related to the cheminformatics similarity searching methods are machine learning methods, such as artificial neural networks, Kohonen self-organizing maps, Naïve Bayes classifiers, and SVMs (support vector machines), which try to align the chemical and biological spaces based on mapping procedures (Savchuck et al. 2004). The goal here is to identify which parts (islands) of the chemical-property space correspond to specific target family or therapeutic activities, and vice-versa. A number of groups have applied such methods to design broad GPCR-focused libraries, and more recently, to specifically distinguish class A, B, and C family subgroup GPCR ligands, or to identify specific GPCR ligands for the adenosine A2A, cannabinoid, CRF, and endothelin GPCRs (von Kroff and Steger 2004).

During the next few years, our detailed knowledge of many newly deorphanized GPCRs and the organization and regulation of the networks of interacting proteins making up the receptosoms will grow. In the perspective of drug discovery, it will be especially interesting to follow how signaling drugs will be discovered further downstream, or whether the GPCR ligand-binding sites will stay the preferred entry point for medication.

Acknowledgements. Drs. R. Bouhelal, M. Gerspacher, and K. Seuwen (all NIBR associates) are acknowledged for various discussions. The text of this chapter is based in major parts of our previous publications.

References

Bemis GW, Murcko MA (1996) The properties of known drugs. 1. Molecular frameworks. J Med Chem 39:2887–2893

Bleicher KH, Green LG, Martin RE, Rogers-Evans M (2004) Ligand identification for G-protein-coupled receptors: a lead generation perspective. Curr Opin Chem Biol 8:287–296

Crossley R (2004) The design of screening libraries targeted at G-protein coupled receptors. Curr Top Med Chem 4:581–588

Evans BE, Rittle KE, Bock MG, DiPardo RM, Freidinger RM, Whitter WL, Lundell GF, Veber DF, Anderson PS, Chang RS (1988) Methods for drug discovery: development of potent, selective, orally effective cholecystokinin antagonists. J Med Chem 31:2235–2246

Guo T, Hobbs DW (2003) Privileged structure-based combinatorial libraries targeting G protein-coupled receptors. Assay Drug Dev Technol 1:579–592

Halazy S (1999) G-protein coupled receptors bivalent ligands and drug design. Exp Opin Ther Patents 9:431–446

Hruby VJ (2002) Designing peptide receptor agonists and antagonists. Nat Rev Drug Discov 1:847–858

Jacoby E, Fauchère JL, Raimbaud E, Ollivier S, Michel A, Spedding M (1999) A three binding site hypothesis for the interaction of ligands with monoamine G-protein coupled receptors: implications for combinatorial ligand design. Quant Struct Act Relat 18:561–572

Jacoby E (2001) A novel chemogenomics knowledge-based ligand design strategy – application to G-protein coupled receptors. Quant Struct Act Relat 20:115–123

Jacoby E, Bouhelal R, Gerspacher M, Seuwen K (2006) The 7TM G-protein-coupled receptor target family. Chem Med Chem 1:760–782

Jimonet P, Jäger R (2004) Strategies for designing GPCR-focused libraries and screening sets. Curr Opin Drug Discovery Dev 7:325–333

Klabunde T, Hessler G (2002) Drug design strategies for targeting G-protein-coupled receptors. Chem Bio Chem 3:928–944

Savchuck NP, Balakin KV, Tkachenko SE (2004) Exploring the chemogenomic knowledge space with annotated chemical libraries. Curr Opin Chem Biol 8:412–417

Schuffenhauer A, Floersheim P, Acklin P, Jacoby E (2003) Similarity metrics for ligands reflecting the similarity of the target proteins. J Chem Inf Comput Sci 43:391–405

Tyndall JDA, Pfeiffer B, Abbenante G, Fairlie DP (2005) Over one hundred peptide-activated G protein-coupled receptors recognize ligands with turn structure. Chem Rev 105:793–826

Von Korff M, Steger M (2004) GPCR-tailored pharmacophore pattern recognition of small molecular ligands. J Chem Inf Comput Sci 44:1137–1147

Webb TR (2004) Some principles related to chemogenomics in compound library and template design for GPCRs. In: Kubinyi H, Müller G (eds) Chemogenomics in drug discovery – a medicinal chemistry perspective. Wiley, Weinheim, pp 313–324

Wermuth CG (2004) Selective optimization of side activities: another way for drug discovery. J Med Chem 47:1303–1314

Wise A, Jupe SC, Rees S (2004) The identification of ligands at orphan G-protein coupled receptors. Annu Rev Pharmacol Toxicol 44:43–66

Ernst Schering Foundation Symposium Proceedings, Vol. 2, pp. 105–143
DOI 10.1007/2789_2006_003

Published Online: 16 May 2007

Orphan Seven Transmembrane Receptor Screening

M.J. Wigglesworth(✉), L.A. Wolfe, A. Wise

Screening and Compound Profiling, GlaxoSmithKline, New Frontiers Science Park, Third Avenue, Harlow, CM19 5AW Essex, UK
email: *Mark.J.Wigglesworth@gsk.com*

Abstract. Drug discovery has successfully exploited the superfamily of seven transmembrane receptors (7TMR), with over 35% of clinically marketed drugs targeting them. However, it is clear that there remains an undefined potential within this protein family for successful drugs of the future. The human genome sequencing project identified approximately 720 genes that belong to the 7TMR superfamily. Around half of these genes encode sensory receptors, while the other half are potential drug targets. Natural ligands have been identified for approximately 215 of these, leaving 155 receptors classified as orphan 7TMRs having no known ligand. Deorphanisation of these receptors by identification of natural ligands has been the traditional method enabling target validation by use of these ligands as tools to define biological relevance and disease association. Such ligands have been paired with their cognate receptor experimentally by screening of small molecule and peptide ligands, reverse pharmacology and the use of bioinformatics to predict candidate ligands. In this manuscript, we review the methodologies developed for the identification of ligands at orphan 7TMRs and exemplify these with case studies.

1 Introduction

The seven transmembrane receptors (7TMRs) have diverse functional roles epitomised by the pharmacological agents of marked therapeutic benefit that target them. Of the approximately 500 clinically marketed drugs greater than 35% are modulators of 7TMR function. This represents more than US $30 billion dollars worth of pharmaceutical sales, making 7TMRs the most successful of any target class in terms of drug discovery (Drews 2000). Yet these drugs exert their activity at less than 10% of cloned 7TMRs, by targeting approximately 30 well-characterised receptors. The human genome sequencing project identified approximately 720 genes that belong to the 7TMR superfamily (Venter et al. 2001; Lander et al. 2001). All of these proteins possess a domain containing seven membrane spanning helixes, a putative extracellular ligand-binding domain and an intracellular domain accessible to G-proteins or other intracellular signalling proteins. It is predicted that half of these genes encode sensory receptors, while the other half are potential drug targets. The natural ligand has been identified for around 215 of these, leaving 155 receptors classified as orphan 7TMRs, having no known ligand and with often little known about their function. Traditionally target validation has relied upon considerable resource to deorphanise these receptors, enabling the use of their natural ligands as tools to define physiological roles and disease association. Using similar methods, surrogate ligands such as small drug-like molecules, toxins and synthetic peptides can also be identified. Examples of natural and surrogate ligands can be found in Table 1. The basis of success-

Table 1 Ligand-receptor pairings from 1993 on

Receptor	Assay	Ligand	Ligand source	References
BRS-3	Oocytes	Bombesin	Surrogate synthetic	Fathi et al. 1993
FPRL-1	Radioligand binding cAMP	Lipoxin A4	Known synthetic	Fiore et al. 1994
ORL-1	cAMP	Nociceptin/ orphanin FQ	Novel from brain extract	Meunier et al. 1995
AZ3B	$[Ca^{2+}]_i$, oocytes	C3a	Known synthetic	Ames et al. 1996
CIRL	Radioligand binding	Latrotoxin	Surrogate synthetic	Krasnoperov et al. 1997
APJ	Ext. pH	Apelin	Novel from stomach extract	Tatemoto et al. 1998
CRLR	Oocytes	CGRP	Known synthetic	McLatchie et al. 1998
GPR10	Arachidonic acid	Prolactin-releasing peptide	Novel from brain extract	Hinuma et al. 1998
Orexin-1 and 2	$[Ca^{2+}]_I$	Orexin-A and B	Novel from brain extract	Sakurai et al. 1998; De Lecea et al. 1998
EDG1, 3, 5, 6 and 8	$[Ca^{2+}]_i$, cAMP	S1P	Known synthetic	An et al. 1997; Lee et al. 1998; Yamazaki et al. 2000; Im et al. 2000a
EDG2, 4 and 7	$[Ca^{2+}]_i$, cAMP	LPA	Known synthetic	Erickson et al. 1998; An et al. 1998; Im et al. 2000b
GHS-R	$[Ca^{2+}]_I$	Ghrelin	Novel from stomach extract	Kojima et al. 1999
FPRL-1	$[Ca^{2+}]_I$	Serum amyloid A	Known synthetic	Su et al. 1999

Table 1 (continued)

Receptor	Assay	Ligand	Ligand source	References
GPR14	$[Ca^{2+}]_I$	Urotensin II	Novel from brain extract and synthetic	Mori et al. 1999; Ames et al. 1999
GPR38	$[Ca^{2+}]_I$	Motilin	Known synthetic	Feighner et al. 1999
HG57 (Cys-LT1R)	$[Ca^{2+}]_I$	LTD4	Known synthetic	Lynch et al. 1999
SLC-1 (MCH1)	$[Ca^{2+}]_I$	Melanin-concentrating hormone	Known frombrain extract and synthetic	Chambers et al. 1999; Lembo et al. 1999
CRTH2	$[Ca^{2+}]_I$	Prostaglandin D2	Known synthetic	Hirai et al. 2001
FM-3/4	$[Ca^{2+}]_I$	Neuromedin U	Known synthetic	Nothacker et al. 1999; Howard et al. 2000; Kojima et al. 2000; Szekeres et al. 2000
GPR16 (BLT2)	cAMP, radioligand binding	LTB4	Known synthetic	Kamohara et al. 2000
GPRv53 (H4-R)	cAMP, radioligand binding	Histamine	Known synthetic	Zhuet al. 2001b; Nakamura et al. 2000; Morse et al. 2001; Nguyen et al. 2001; Liu et al. 2001
HLWAR77	$[Ca^{2+}]_I$	Neuropeptides FF and AF	Known synthetic	Elshourbagy et al. 2000
KIAA0001	Yeast	UDP-glucose	Novel synthetic	Chambers et al. 2000
OGR-1	$[Ca^{2+}]_I$	Sphingosylphosphorylcholine	Known synthetic	Xu et al. 2000

Table 1 (continued)

Receptor	Assay	Ligand	Ligand source	References
PSECO146 (CysLT2R)	$[Ca^{2+}]_i$, oocytes	LTC4 and D4	Known synthetic	Heise et al. 2000; Nothacker et al. 2000
AXOR12 (GPR54)	$[Ca^{2+}]_I$	KiSS-1	Novel from placental extract and synthetic	Ohtaki et al. 2001
G2A	$[Ca^{2+}]_I$	Lysophosphatidylcholine	Novel synthetic	Kabarowski et al. 2001
GPR4	$[Ca^{2+}]_I$	Sphingosylphosphorylcholine, lysophosphatidylcholine	Known synthetic	Zhu et al. 2001a
MCH2	$[Ca^{2+}]_I$	Melanin-concentrating hormone	Known synthetic	Hill et al. 2001
P2Y12	Oocytes	ADP	Known synthetic	Hollopeter et al. 2001
TA_1, TA_2	Oocytes	Trace amines (tyramine	Known synthetic	Borowsky et al. 2001
TDAG-8	cAMP	Psychosine	Known synthetic	Im et al. 2001
MRGX1	$[Ca^{2+}]_I$	Bovine adrenal medulla 22	Known/surrogate synthetic	Lembo et al. 2002
LGR7 and 8	cAMP	Relaxin	Known synthetic	Hsu et al. 2002
TG1019	TG1019-G_{i1}a fusion, GTPgS binding	Eicosanoids, unsaturated fatty acids	Known synthetic	Hosoi et al. 2002; Jones et al. 2003
BG37/TGR5	cAMP	Bile acids	Known synthetic	Maruyama et al. 2002; Kawamata et al. 2003
C5L2 (GPR77)	Radioligand binding	C5a, C5a des Arg, C3a des Arg (acylation-stimulating protein)	Known synthetic	Cain and Monk 2002; Kalant et al. 2003
GPR7 and 8	cAMP	Neuropeptides B and W	Novel from brain extract and synthetic	Shimomura et al. 2002; Fujii et al. 2002; Brezillon et al. 2003

Table 1 (continued)

Receptor	Assay	Ligand	Ligand source	References
ChemR23	$[Ca^{2+}]_I$	Tazarotene-induced gene-2 (chemerin)	Novel synthetic	Meder et al. 2003
GPR103 (AQ27)	$[Ca^{2+}]_I$	QRFP	Novel synthetic	Fukusumi et al. 2003
MRGX2	$[Ca^{2+}]_I$ cAMP	Cortistatin	Known synthetic	Robas et al. 2003
		PAMP	Known synthetic	Kamohara et al. 2005
GPR40	$[Ca^{2+}]_I$	Medium/long chain fatty acids	Novel synthetic	Briscoe et al. 2003; Brown et al. 2003; Itoh et al. 2003; Kotarsky et al. 2003
GPR41 and 43	Yeast	Short chain fatty acids	Novel synthetic	Brown et al. 2003
HM74A	GTPgS binding	Nicotinic acid	Known synthetic	Soga et al. 2003; Tunaru et al. 2003; Wise et al. 2003
GPR30	cAMP ERK	17 β-Estradiol	Known synthetic	Maggiolini et al. 2004; Kanda and Watanabe 2003
FPR, FPRL-1, FPRL-2	$[Ca^{2+}]_I$	Annexin-1	Known synthetic	Ernst et al. 2004
FPRL-1	$[Ca^{2+}]_I$	CKbeta8-1	Known synthetic	Elagoz et al. 2004
FPRL-1 and 2	Chemotaxis, $[Ca^{2+}]_i$	Humanin	Known from stomach extract	Ying et al. 2004; Harada et al. 2004

Table 1 (continued)

Receptor	Assay	Ligand	Ligand source	References
GPR80 P2Y15	$[Ca^{2+}]_i$, cAMP radioligand binding	Alpha-ketoglutarate AMP and adenosine	Known from kidney extracts	He et al. 2004; Inbe et al. 2004
GPR91	$[Ca^{2+}]_i$, cAMP	Succinate	Known synthetic	He et al. 2004
NPS	$[Ca^{2+}]_I$	Neuropeptide S	Novel synthetic	Xu et al. 2004
TDAG8	cAMP	Acid-sensing	Novel synthetic	Ishii et al. 2005
CXCR7 (RDC1)	Radioligand-binding and chemotaxis	CXCL12	Known synthetic	Balabanian et al. 2005
GPR39	cAMP	Obestatin	Novel synthetic	Zhang et al. 2005a
GPR120	$[Ca^{2+}]_I$ERK	Fatty acids	Known synthetic	Hirasawa et al. 2005
GPRC6A	Oocytes	L-alpha-amino acid receptor	Surrogate synthetic	Wellendorph et al. 2005
GPR119	Yeast, cAMP	Oleoylethanolamide	Known from human hemofiltrate peptide library	Overton et al. 2006
GPR37	Oocytes, $[Ca^{2+}]_I$	Neuropeptide head activator (HA)	Known synthetic	Rezgaoui et al. 2006

2 Generation of Orphan 7TMR Functional Assays

In past reviews, we have explained the importance of confirming cell-surface receptor expression (Wise et al. 2002, 2004). Naturally, this remains an important aspect of generating a functional assay system. In our experience there are a great many orphan 7TMRs that do not clearly express on the cell surface. This may be due in part to overexpression within the recombinant systems, lack of appropriate trafficking partners or simply due to a natural intracellular localisation of certain orphan 7TMRs. The mechanism of expression can also significantly affect the pattern and amount of expression, as well as the functional response. It has long been accepted that cell lines can be generated with differing expression levels of recombinant receptors. However, the recent use of baculovirus expression within mammalian cells allows scientists to titrate pharmacology by altering viral load. Such technology enables the ability to generate both high- and low-expression screening systems in a multitude of cell types (Boyce and Bucher 1996; Condreay et al. 1999; Ames et al. 2004; Hsu et al. 2004; Kost et al. 2005). Expression within different systems and/or cell types increases the chance that any critical expression partners will be available to enhance cell surface expression and signalling.

One protein that has an interaction with 7TMRs that is useful in characterising their activation is β-arrestin. Measuring the recruitment of β-arrestins provides a means of measuring the internalisation of 7TMRs (Barak et al. 1997; Bertrand et al. 2002; Vrecl et al. 2004). This interaction does not rely upon measuring G-protein activation and in some cases generates a mechanism of measuring novel 7TMR pathways (Oakley et al. 2002; Shenoy and Lefkowitz 2005; Shenoy et al. 2006). Following receptor activation and phosphorylation, β-arrestin is recruited to the cell surface and internalised within endosomes. Chimeric β-arrestin-GFP (green fluorescent protein) can easily be tracked within the cell and upon recruitment to activated receptors. This characteristic compartmentalisation of the β-arrestin-GFP thus indicates the activation of an expressed 7TMR. This technique has been further developed to detect bioluminescent energy transfer (BRET) between 7TMRs and arrestins, when each is a chimera with luminescent proteins. Hence, measuring specific receptor activation as such can be used to measure receptor–ligand interactions (Bertrand et al. 2002; Vrecl et al. 2004; Hamdan et al. 2005). This technique has potential as an assay for deorphanisation of 7TMRs, and has been used to aid characterisation of recently deorphanised receptors (Evans et al. 2001). However, few ligand–receptor pairings have utilised this technology, one example in the literature being the deorphanisation of Drosophila neuropeptide receptors (Johnson et al. 2003). Although we should also note the recruitment of β-arrestin is a characteristic shared by many 7TMRs, there are a small number that do not interact with β-arrestin (Perroy et al. 2003; Breit et al. 2004).

There are a number of generic assay systems that monitor a wide range of known G-protein-linked effectors. The following assays have often been the first approach to deorphanisation.

The fluorescent imaging plate reader (FLIPR; Molecular Devices, Sunnyvale, CA, USA), utilises calcium-sensitive dyes in combination with promiscuous G-proteins such as $G\alpha_{16/15}$ and chimeric G-proteins such as $G\alpha_{qs5}$, $G\alpha_{si5}$ and $G\alpha_{16z49}$ (Offermanns and Simon 1995; Mody et al. 2000; Milligan 2000; Kostenis 2001). These G-proteins are combined in transient expression within mammalian cells, facilitating coupling of a wide range of receptors to the $G\alpha_q$ pathway. This ultimately allows receptor activation to be monitored as calcium mobilisation downstream.

Adaptation of the pheromone receptor pathway (Ste2p and Ste3p) within the fission yeast *Saccharomyces cerevisiae* allows generation of a second generic assay (Blumer and Thorner 1991; Broach and Thorner 1996; Schrick et al. 1997). Chimeric G-proteins are again used; however, in this case they are all enabling the yeast G-protein Gpa1-linked signalling via a MAPKinase cascade. The yeast system provides an excellent basis for screening given its ease of application, cost and low endogenous 7TMR background. There is only one other 7TMR known to be expressed in *Saccharomyces cerevisiae*: a nutrient-sensing receptor GPR1 (Xue et al. 1998; Kraakman et al. 1999; Maidan et al. 2005). Additionally, elevated baseline activity can be used to identify recombinant orphan 7TMRs that induce constitutive activity. This is useful in indicating the likely G-protein coupling specificity of the receptor (Medici et al. 1997; Dowell and Brown 2002; Pausch et al. 2004).

Xenopus melanophores provide an assay system that cleverly exploits nature (Graminski et al. 1993; McClintock et al. 1993; Lerner

1994). These cells are from the neural crest of *Xenopus laevis* and contain the dark brown pigment melanin within intracellular organelles called melanosomes. A 7TMR that is expressed and couples to G-protein will regulate the translocation of these melanosomes causing dispersion or aggregation. Receptor expression can be titrated by use of various cDNA amounts, and as this expression alters we can monitor constitutive G-protein coupling. This has proven to be a very powerful tool in predicting G-protein activation, with the vast majority of 7TMRs showing activation in this system when overexpressed. Constitutive or ligand activated pigment dispersion is the signature of adenylate cyclase or phospholipase C activation, whereas aggregation signifies inhibition of adenylate cyclase (Chen et al. 1999). When pigment is dispersed, light transmission is reduced through the cell, and when cells are in the aggregated state, the pigment collects around the nucleus, resulting in increased light transmission. This provides a rapid and very sensitive screening system, with potency of known ligands often being increased within a melanophore system compared to traditional mammalian assays.

Other assays that are utilised include cAMP assays for $G\alpha_i$ and $G\alpha_s$-coupled receptors (Hemmila 1999; Eglen 2005), GTPγS mainly for $G\alpha_i$ (Milligan 2003), FLIPR or Aequorin, a reporter gene-based calcium assay, for $G\alpha_q$ (Stables et al. 1997), while other reporter gene assays can be utilised for a wide array of specific pathways (Goetz et al. 1999; Durocher et al. 2000; Rees et al. 2001; Kunapuli et al. 2003). The $G\alpha_{12}$ and $G\alpha_{13}$ G-protein subfamily represent the last well-characterised family of G-proteins. There are some receptors that have been shown to couple to these G-proteins, members of the EDG family of receptors and Galanin receptors, for example (Windh et al. 1999; Wittau et al. 2000). It is of course possible that a number of orphan 7TMRs couple and signal via $G\alpha_{12}$ and $G\alpha_{13}$ and hence the generation of convenient high-throughput screening systems based on activation of this G-protein subfamily may prove beneficial for future orphan 7TMR screening strategies.

3 Native or Surrogate Ligands?

A pragmatist in the current environment of orphan 7TMR research may consider deorphanisation a difficult and unnecessary approach. Definition of targets and disease association is possible by detailed localisation studies and widespread access to genetic linkage data. Surrogate ligands can be identified by expressing and screening the 7TMR against libraries of diverse small drug-like molecules, within the generic assay systems above. These tools can then be used to validate disease association via specific physiological assays in much the same way as the native ligand. However, progressing targets for which we have surrogate ligands but no natural ligand identified can be a risky strategy, as there is no way of knowing whether such molecules bind to orthosteric or allosteric sites. The emerging phenomena of ligand-directed coupling, for example (Heusler et al. 2005; Simmons 2005), suggests that ligands can be identified that show dual or selective G-protein activation. In these cases, identification of the signalling pathway most relevant to the physiological end point is paramount to the success of any subsequent drug discovery progression path. The use of surrogate ligands to enable screens for antagonist molecules may also prove risky, as there is no guarantee that blocking a surrogate agonist would also block the effects of the endogenous ligand. Thus, native ligand pairings remain the primary way of evoking target validation and effective prosecution of a drug discovery process.

4 Candidate Native Ligand Identification

Compilation of candidate native ligand sets has been ongoing within pharmaceutical companies for many years. Obvious additions are known 7TMR ligands: small molecules such as histamine, lipids such as sphingosylphosphorylcholine and peptides, for example motilin and ghrelin. Identification of novel ligands relies on known biological function, similarity to known ligands and predictions made from bioinformatics analysis. Ligand sets from each of these sources are now available as plates ready for screening from various commercial sources. An alternative approach is to search for activation of your target 7TMR by tissue extracts. Activity of specific extracts signifies the existence of a native ligand, demonstrates that your screening system is functional and may suggest specific tissues where a 7TMR will have biological effects. However, identifying the active component within the extract remains a substantial obstacle. While this approach has been successful (Habata et al. 1999; Kojima et al. 1999; Tanaka et al. 2003), there are many technical challenges that have to be addressed to ensure successful pairing, and it is a highly resource-intensive approach.

5 Bead-Based Screening Within *Xenopus melanophores*

Peptide receptors have long been considered good drug targets due to their association with disease. PTH-1, MC4R, and GLP-1 are just a few of the many examples. Orphan 7TMRs that are capable of responding to peptide ligands may also play a key role in disease and one day might be deemed desirable drug targets. By utilizing the power of combinatorial chemistry to generate large synthetic peptide libraries, together with the robustness of melanophores in cell-based assays, millions of peptides can be screened functionally and individually against dozens of orphan receptors in an effort to pair peptides to receptors.

Jayawickreme et al. (1999) previously demonstrated that one could combine the technologies of combinatorial chemistry and melanophore assays to allow the rapid screening of a 442,368-member peptide library as discrete molecules in a lawn format. To expand on this approach, a larger bead-based photo-cleavable library was generated and screened against melanophores transiently expressing multiple peptide-like receptors simultaneously. This library was constructed from 17 L- and 5 D-amino acids resulting in 22^6, or 113,379,904 unique peptides. To make screening manageable, two decisions were made. First, beads were pooled so that each pool would contain the same first amino acid. Therefore, only 22 total pools needed screening. Second, assuming related peptides would likely behave somewhat similarly, only a one-tenth equivalent (11.3 M) of the library was screened to determine if active peptides were present.

The cell lawn contains melanophores transiently transfected with peptide multiple receptor (PMR) sets. Each PMR set contains a mixture of known liganded peptide receptors and orphan receptors predicted to have peptide ligands. Since pigment translocation in melanophores is dependent on G-protein coupling, PMR sets contain either $G\alpha_s/G\alpha_q$-coupled or $G\alpha_i/G\alpha_o$-coupled receptors, but not both. Peptide beads are suspended in agar and layered above the cell lawn, exposed to UV light, and those that cause pigment translocation are selected. Each active bead is then retested to confirm the peptide's specificity to the PMR set. Beads containing peptides with continued interest are then sequenced and peptides synthesised. Receptor deconvolution and evaluation of peptide activity can then be achieved by screening all receptors from the PMR set individually against the synthetic peptide(s) in a dose-response manner in a typical well-based assay system. Ligands like these could then be used as tools for in vivo and in vitro target validation studies, high throughput screening tools for drug discovery, for modelling into small molecule drugs, or for the generation and screening of new focused libraries.

6 Knowledge-Based Deorphanisation Approaches

The 7TMR phylogeny demonstrates clearly that family members with high sequence similarity respond to the same ligands. Deorphanisation of a single member of a receptor family has often led to successful pairings of the entire cluster of receptors. It is easy to forget in a post-human-genome era that this was not always this obvious. Pairings of this sort have occurred over time as the novel receptor sequences became available. The cloning of an expressed sequence tag bearing significant similarity to biogenic amine receptors including the H3 receptor, resulted in the pairing of histamine with a fourth histamine receptor (Oda et al. 2000; Nakamura et al. 2000; Morse et al. 2001; Nguyen et al. 2001).

Ligand–receptor pairing by logical application of data is perhaps best exemplified by deorphanisation of HM74 (Wise et al. 2003). It was known that nicotinic acid was clinically effective in reducing a range of markers for cardiovascular disease. Its effects include a reduction in high-density lipoprotein and mortality rates and normalisation of a range of cardiovascular risk factors in patient populations. It was also believed that nicotinic acid caused activation of a $G\alpha_i$-linked G-protein-coupled receptor in adipocytes (Green et al. 1992). This matched evidence that the clinically relevant site of action of nicotinic acid was restricted to adipose and spleen tissue (Lorenzen et al. 2001). However, the mechanism by which nicotinic acid produced these desirable effects still needed to be elucidated.

A list of ten candidate orphan GPCRs was identified by selecting orphan receptors with mRNA tissue distribution patterns that correlated with known pharmacological sites of action of niacin, e.g. adipose and spleen. These candidate receptors were then screened in a GTPγS assay to measure activation of Gα_i-G proteins upon exposure to nicotinic acid. An orphan 7TMR HM74 demonstrated low-potency responses to nicotinic acid in this assay that could not solely account for high-potency tissue responses. Bioinformatics analysis identified HM74A and GPR81 with 96% and 57% identity to HM74, respectively, as possible family members that could share the same ligand. Both receptors were also restricted to adipose and spleen tissues by mRNA expression profiling. HM74A was subsequently identified as a high-affinity receptor for nicotinic acid (Wise et al. 2003). A number of similarly interesting ligands have been described in Table 2. It will be interesting to see in the future if similar knowledge-based approaches will aid in the identification of novel ligand–receptor pairings.

7 Protein Collaboration in the 7TMR World

Over recent years, it has become clear that some 7TMRs require secondary proteins in order to function. This can be due to a total loss of cell surface expression, as is the case for the GABA-B-R1 receptor, which requires co-expression of GABA-B-R2 as a trafficking protein to deliver the functional receptor to the cell surface (White et al. 1998; Kaupmann et al. 1998; Jones et al. 1998). Opioid receptors have also been shown to collaborate with each other: co-expression of these receptors greatly alters pharmacology (Jordan and Devi 1999; Gomes et al. 2000, 2004; George et al. 2000). It is clear that the number of 7TMRs shown to dimerise is rapidly growing (White et al. 1998; George et al. 2000; Rocheville et al. 2000; Lee et al. 2000; Breit et al. 2004; Gomes et al. 2004; Breitwieser 2004; Zhu et al. 2005; O'Dowd et al.

2005). It is possible that some of the remaining orphan receptors are expression partners or that they require expression partners in order to produce a functional receptor. However, although there are examples, functional effects of dimerisation have rarely been measured in such ways, perhaps further implying the difficulties in the field of both dimerisation and orphan 7TMRs.

7TMRs are not family monogamous in their protein interactions. There are many characterised proteins known to interact directly or indirectly with 7TMRs (Bockaert et al. 2004). One example that had the potential to expand the possibilities of ligand pairing is the family of single transmembrane-spanning proteins known as receptor activity modifying proteins (RAMPs). They are known to interact with a small number of receptors and were first characterised by their ability to direct the pharmacology of the calcitonin receptor-like receptor (CRLR). Calcitonin gene-related peptide (CGRP) and adrenomedullin receptors are created when co-expressed with RAMP-1, or RAMP-2 and RAMP-3, respectively (McLatchie et al. 1998; Christopoulos et al. 2003).

8 TMRs but Not GPCRs

The C5L2, C5a and C3a family of 7TMRs have been shown to have varying affinity for C5a, C3a and their desarginated forms C5adR^{74} and C3adR^{77}, respectively (Cain and Monk 2002; Kalant et al. 2003, 2005). The C3a and C5a receptors have clear G-protein-mediated effects. However, for C5L2 initial studies failed to find any G-protein-mediated functional response to ligand binding using standard G-protein-dependent second messenger pathways (Okinaga et al. 2003). It was speculated that this receptor may act as a ligand sink, binding to the ligands of other 7TMRs, competing with them for the available ligand and hence reducing the active receptor–ligand complex. A recent publication indicates that in mice that contain a targeted disruption of the C5L2 receptor, the biological activity of C5a is increased (Gerard et al. 2005), demonstrating a biological function for the C5L2 receptor. Further to this, C5L2 has been reported to mediate triglyceride stimulation and cause β-arrestin recruitment (Kalant et al. 2005). More recently, acylation-stimulating protein (ASP) activation of C5L2 has been linked with pathways signifying Gα_q signalling within 3T3-L1 preadipocyte cells (Maslowska et al. 2006). However, in agreement with other literature all our efforts to characterise this receptor within recombinant cell lines have failed (data not shown). It is possible that the expression system is crucial to function and that although ASP and C5L2 are required for the Gα_q signalling observed by Maslowska et al. (2006), these may not be the only requirements. It is possible that the preadipocyte cells used contain an expression partner that enables coupling in much the same manner as the GABA-B receptor (White et al. 1998) or that some other as yet uncharacterised pathway is activated. The difficulties encountered generating functional recombinant assays for receptors such as C5L2 highlights that the classical approaches described above may not be relevant in detecting function of some of the remaining orphan 7TMRs. A conventional functional assay screening approach would not have detected the binding of these ligands to C5L2.

The adiponectin receptors are predicted to possess 7TMRs; however, they have not been shown to be G-protein-coupled (Yamauchi et al. 2003). Both receptors and ligand have been linked to clinical aetiology of liver disease (Jonsson et al. 2005), whilst knockout mice link adiponectin with insulin control (Maeda et al. 2002). They are known to mediate their effects via MAP kinase cascades, although the mechanisms by which they activate this pathway are not yet clear (Luo et al. 2005).

9 Ligandless 7TMRs

The Kaposi's sarcoma-associated herpesvirus encodes ORF74, which is one example of a number of viral 7TMRs that have no endogenous ligand yet display high levels of constitutive activity. ORF74 has evolved to function in the absence of added agonist (Smit et al. 2002; Vischer et al. 2006). Viral 7TMRs have a strong rationale for having no endogenous ligand, given that they must function in an infected host cell. However, it is possible that examples of mammalian orphan 7TMRs that demonstrate similar ligandless mechanisms of action will be described in the future. It is perhaps also worth considering that some receptors have their ligands in situ without obvious functional activity and are then able to detect changes in environmental conditions such as light, as is the case for the epiphany of 7TMRs: rhodopsin. It is also possible that there are a number of orphan 7TMRs waiting for natural ligands to evolve: in evolution what comes first, ligand or receptor?

10 Difficulties in Replicating Literature Pairings

Zhang et al. (2005a) have recently characterised obestatin as a novel bioactive peptide ligand for GPR39. Obestatin is derived from the same gene as ghrelin, and interestingly reverses the increased appetite effects of ghrelin. Within GlaxoSmithKline, we are able to reproduce the in vivo and in vitro tissue effects of obestatin (G. Sanger and A. Bassil, personal communication). However, thus far we have been unable to reproduce the pairing of GPR39 with obestatin.

GPR39 has been profiled at GlaxoSmithKline through a variety of ligand screening assays we have discussed above. By virtue of a FLIPR reverse pharmacology approach, we have identified nickel chloride as a surrogate ligand for this receptor (data not shown), and additionally in melanophore bead-based screening experiments, we paired two surrogate hexamer peptides. These peptides demonstrate potency of 6.77 and 4.73 at GPR39 when recombinantly expressed within the melanophore system, and activity has been confirmed within mammalian FLIPR assays where Gq coupling is demonstrated. Using these peptides as surrogate ligands to demonstrate functional expression of GPR39 in our assay systems we have, however, been unable to replicate the pairing of GPR39 with obestatin. This perhaps indicates that the pharmacology of GPR39 is somewhat complex. It may be regulated by expression partners absent from our cell lines or it may be able to traffic its responses via different G-proteins depending on the interacting ligand.

11 The MrgX Paradigm

The MrgX family of receptors is the perfect example of why we have struggled to identify natural ligands for all 7TMRs. Human MrgX receptors are also known as sensory neuron specific receptors, and as the name suggests, they are specifically expressed within primary afferents and dorsal root ganglion (Dong et al. 2001). As such they are potential pain-modulating 7TMRs and of interest to several pharmaceutical companies. Within this family, we find a large number of pitfalls for orphan biology.

The MrgX family of receptors contain nine closely related receptors: MrgX1–7, MrgD and MrgE (Zhang et al. 2005b). The overall sequence identity of these receptors would indicate that they could bind and be activated by similar if not the same ligands. Conversely, detailed evolutionary analysis suggests that the ligand-binding region of this family across species has undergone strong selection to diversify their ligand-binding properties (Choi and Lahn 2003). In agreement with this, these receptors appear to have very diverse ligands (Lembo et al. 2002; Han et al. 2002; Zylka et al. 2003; Robas et al. 2003; Shinohara et al. 2004). Two of the human receptors, MrgX1 and MrgX2, bind bovine adrenal medulla (BAM) peptides and cortistatin, respectively (Lembo et al. 2002; Robas et al. 2003). MrgX2 appears to be promiscuous and can be activated by several peptides at lower potency than cortistatin (Robas et al. 2003; Kamohara et al. 2005). MrgX3 and MrgX4 remain orphan receptors; however, within GlaxoSmithKline we have identified surrogate small molecule ligands for MrgX4, demonstrating that if the MrgX4 receptor were activated by the same peptide ligands as MrgX1 or MrgX2, we should have detected this within our assays. Further to the difficulties of this family, we have thus far failed to find any cell surface expression, ligands or function for MrgX3.

The BAM peptides *in vivo* rat models are pronociceptive, providing positive pain validation for the peptide ligand (Grazzini et al. 2004). Recombinant expression of MrgX1 within cultured rat neurons produces signalling classical of painful stimuli (Chen and Ikeda 2004). Cortistatin has also been linked with disorders of sleep, locomotor and cortical function. Yet, in rodents the MrgX receptor story becomes much more complex. There are in the region of 30 rat receptors and even more mouse receptors that have similarity to the human MrgX family (Lembo et al. 2002; Zylka et al. 2003; Burstein et al. 2006). Conversely to literature publication, the human and rodent MrgX receptors are not similar enough to identify direct orthologues, although functional equivalence has been identified with MrgX1 (Lembo et al. 2002). However, rSNSR can also be activated with high potency by γ2-MSH among other ligands (Grazzini et al. 2004), suggesting that this is a promiscuous receptor in terms of its ligand interactions.

Within the public databases, there is more than one sequence reported for the rSNSR receptor. Initially the receptor used at GlaxoSmithKline, although identical to one published sequence, was different to the sequence used in the initial pairing publication (Lembo et al. 2002) by three seemingly innocuous amino acids. However, our sequence failed to express in mammalian transient systems, while melanophore and yeast assays indicated only low-potency activity. Hence, we failed to reproduce the ligand pairing. Only when the rodent receptor was mutated back to the publication sequence did we restore activity that matched the publication. In this case, it appears that polymorphisms within the genome of rodent receptors can also alter the function of 7TMRs and prevent ligand pairing. This work highlights the importance of using the same receptor splice variant or polymorph for ligand confirmations.

In addition, a recent publication demonstrates the functional effects of co-expressing MRGD and MRGE receptors (Milasta et al. 2006). In situ hybridisation also suggests that these receptors co-express in vivo, possibly indicating a functional significance (Zylka et al. 2003). It would be interesting to see if this is also the case for the human receptors with MrgX4 which remain a complete enigma.

Many MrgX receptors demonstrate significant constitutive activity and can couple to both Gq and Gi pathways (Burstein et al. 2006). Indeed inverse agonists have been reported for MrgX2 in GTPγS assays (Takeda et al. 2003). Burstein et al. (2006) also question the relevance of BAM ligands, given that these peptides show reduced potency at the rhesus monkey orthologues. They speculate that there may be no natural ligand for these receptors. Hence, inverse agonism of a constitutive receptor may be relevant for modifying its role in disease pathophysiology.

12 Future Perspectives

Traditional approaches for the pairing of ligands with orphan 7TMRs have been largely based on the screening of putative GPCR ligands.

This strategy has perhaps overfished the orphan 7TMRs oceans, leaving only receptors that will require novel baits and detection methods. In recent years, several technological and knowledge-based advances have been utilised to successfully deorphanise 7TMRs. However, they have yet to alter the rate of deorphanisation reports. Allowing novel concepts to shape the way we approach orphan 7TMRs will be critical in further advancing this field. This may be via identification of novel activation pathways that do not require G-proteins, exploitation of $G\alpha_{12/13}$ assays, developing an understanding of function that does not require a ligand for activation or by identification of novel ligands, both natural and surrogate.

References

Albertin G, Malendowicz LK, Macchi C, Markowska A, Nussdorfer GG (2000) Cerebellin stimulates the secretory activity of the rat adrenal gland: in vitro and in vivo studies. Neuropeptides 34:7–11

Ames RS, Li Y, Sarau HM, Nuthulaganti P, Foley JJ, Ellis C, Zeng Z, Su K, Jurewicz AJ, Hertzberg RP, Bergsma DJ, Kumar C (1996) Molecular cloning and characterization of the human anaphylatoxin C3a receptor. J Biol Chem 271:20231–20234

Ames RS, Sarau HM, Chambers JK, Willette RN, Aiyar NV, Romanic AM, Louden CS, Foley JJ, Sauermelch CF, Coatney RW, Ao Z, Disa J, Holmes SD, Stadel JM, Martin JD, Liu WS, Glover GI, Wilson S, McNulty DE, Ellis CE, Elshourbagy NA, Shabon U, Trill JJ, Hay DW, Ohlstein EH, Bergsma DJ, Douglas SA (1999) Human urotensin-II Is a potent vasoconstrictor and agonist for the orphan receptor GPR. Nature 401:282–286

Ames R, Fornwald J, Nuthulaganti P, Trill J, Foley J, Buckley P, Kost T, Wu Z, Romanos M (2004) BacMam recombinant baculoviruses in G protein-coupled receptor drug discovery. Receptors Channels 10:99–107

An S, Bleu T, Huang W, Hallmark OG, Coughlin SR, Goetzl EJ (1997) Identification of CDNAs encoding two G protein-coupled receptors for lysosphingolipids. FEBS Lett 417:279–282

An S, Bleu T, Hallmark OG, Goetzl EJ (1998) Characterization of a novel subtype of human G protein-coupled receptor for lysophosphatidic acid. J Biol Chem 273:7906–7910

Angelone T, Goumon Y, Cerra MC, Metz-Boutigue MH, Aunis D, Tota B (2006) The emerging cardio-inhibitory role of the hippocampal cholinergic neurostimulating peptide. J Pharmacol Exp Ther 318:336–344

Balabanian K, Lagane B, Infantino S, Chow KY, Harriague J, Moepps B, Arenzana-Seisdedos F, Thelen M, Bachelerie F (2005) The chemokine SDF-1/CXCL12 binds to and signals through the orphan receptor RDC1 in T lymphocytes. J Biol Chem 280:35760–35766

Barak LS, Ferguson SS, Zhang J, Caron MG (1997) A beta-arrestin/green fluorescent protein biosensor for detecting G protein-coupled receptor activation. J Biol Chem 272:27497–27500

Bertrand L, Parent S, Caron M, Legault M, Joly E, Angers S, Bouvier M, Brown M, Houle B, Menard L (2002) The BRET2/arrestin assay in stable recombinant cells: a platform to screen for compounds that interact with G protein-coupled receptors (GPCRS). J Recept Signal Transduct Res 22:533–541

Blumer KJ, Thorner J (1991) Receptor-G protein signaling in yeast. Annu Rev Physiol 53:37–57

Bockaert J, Dumuis A, Fagni L, Marin P (2004) GPCR-GIP networks: a first step in the discovery of new therapeutic drugs? Curr Opin Drug Discov Devel 7:649–657

Borowsky B, Adham N, Jones KA, Raddatz R, Artymyshyn R, Ogozalek KL, Durkin MM, Lakhlani PP, Bonini JA, Pathirana S, Boyle N, Pu X, Kouranova E, Lichtblau H, Ochoa FY, Branchek TA, Gerald C (2001) Trace amines: identification of a family of mammalian G protein-coupled receptors. Proc Natl Acad Sci USA 98:8966–8971

Boyce FM, Bucher NL (1996) Baculovirus-mediated gene transfer into mammalian cells. Proc Natl Acad Sci USA 93:2348–2352

Breit A, Lagace M, Bouvier M (2004) Hetero-oligomerization between beta2- and beta3-adrenergic receptors generates a beta-adrenergic signaling unit with distinct functional properties. J Biol Chem 279:28756–28765

Breitwieser GE (2004) G protein-coupled receptor oligomerization: implications for G Protein activation and cell signaling. Circ Res 94:17–27

Brezillon S, Lannoy V, Franssen JD, Le Poul E, Dupriez V, Lucchetti J, Detheux M, Parmentier M (2003) Identification of natural ligands for the orphan G protein-coupled receptors GPR7 and GPR8. J Biol Chem 278:776–783

Briscoe CP, Tadayyon M, Andrews JL, Benson WG, Chambers JK, Eilert MM, Ellis C, Elshourbagy NA, Goetz AS, Minnick DT, Murdock PR, Sauls HR Jr, Shabon U, Spinage LD, Strum JC, Szekeres PG, Tan KB, Way JM, Ignar DM, Wilson S, Muir AI (2003) The orphan G protein-coupled receptor GPR40 is activated by medium and long chain fatty acids. J Biol Chem 278:11303–11311

Broach JR, Thorner J (1996) High-throughput screening for drug discovery. Nature 384:14–16

Brown AJ, Goldsworthy SM, Barnes AA, Eilert MM, Tcheang L, Daniels D, Muir AI, Wigglesworth MJ, Kinghorn I, Fraser NJ, Pike NB, Strum JC, Steplewski KM, Murdock PR, Holder JC, Marshall FH, Szekeres PG, Wilson S, Ignar DM, Foord SM, Wise A, Dowell SJ (2003) The orphan G protein-coupled receptors GPR41 and GPR43 are activated by propionate and other short chain carboxylic acids. J Biol Chem 278:11312–11319

Burstein ES, Ott TR, Feddock M, Ma JN, Fuhs S, Wong S, Schiffer HH, Brann MR, Nash NR (2006) Characterization of the Mas-related gene family: structural and functional conservation of human and rhesus MrgX receptors. Br J Pharmacol 147:73–82

Cain SA, Monk PN (2002) The orphan receptor C5L2 has high affinity binding sites for complement fragments C5a and C5a Des-Arg(74). J Biol Chem 277:7165–7169

Campana WM, Hiraiwa M, O'Brien JS (1998) Prosaptide activates the MAPK pathway by a G-protein-dependent mechanism essential for enhanced sulfatide synthesis by Schwann cells. FASEB J 12:307–314

Chambers J, Ames RS, Bergsma D, Muir A, Fitzgerald LR, Hervieu G, Dytko GM, Foley JJ, Martin J, Liu WS, Park J, Ellis C, Ganguly S, Konchar S, Cluderay J, Leslie R, Wilson S, Sarau HM (1999) Melanin-concentrating hormone is the cognate ligand for the orphan G-Protein-coupled receptor SLC-1. Nature 400:261–265

Chambers JK, Macdonald LE, Sarau HM, Ames RS, Freeman K, Foley JJ, Zhu Y, McLaughlin MM, Murdock P, McMillan L, Trill J, Swift A, Aiyar N, Taylor P, Vawter L, Naheed S, Szekeres P, Hervieu G, Scott C, Watson JM, Murphy AJ, Duzic E, Klein C, Bergsma DJ, Wilson S, Livi GP (2000) A G protein-coupled receptor for UDP-glucose. J Biol Chem 275:10767–10771

Chen G, Jayawickreme C, Way J, Armour S, Queen K, Watson C, Ignar D, Chen WJ, Kenakin T (1999) Constitutive receptor systems for drug discovery. J Pharmacol Toxicol Methods 42:199–206

Chen H, Ikeda SR (2004) Modulation of ion channels and synaptic transmission by a human sensory neuron-specific G-protein-coupled receptor, SNSR4/MrgX1, heterologously expressed in cultured rat neurons. J Neurosci 24:5044–5053

Choi SS, Lahn BT (2003) Adaptive evolution of MRG, a neuron-specific gene family implicated in nociception. Genome Res 13:2252–2259

Christopoulos A, Christopoulos G, Morfis M, Udawela M, Laburthe M, Couvineau A, Kuwasako K, Tilakaratne N, Sexton PM (2003) Novel receptor partners and function of receptor activity-modifying proteins. J Biol Chem 278:3293–3297

Condreay JP, Witherspoon SM, Clay WC, Kost TA (1999) Transient and stable gene expression in mammalian cells transduced with a recombinant baculovirus vector. Proc Natl Acad Sci USA 96:127–132

De Lecea L, Kilduff TS, Peyron C, Gao X, Foye PE, Danielson PE, Fukuhara C, Battenberg EL, Gautvik VT, Bartlett FS, Frankel WN, van den Pol AN, Bloom FE, Gautvik KM, Sutcliffe JG (1998) The hypocretins: hypothalamus-specific peptides with neuroexcitatory activity. Proc Natl Acad Sci USA 95:322–327

Dong X, Han S, Zylka MJ, Simon MI, Anderson DJ (2001) A diverse family of GPCRs expressed in specific subsets of nociceptive sensory neurons. Cell 106:619–632

Dowell SJ, Brown AJ (2002) Yeast assays for G-protein-coupled receptors. Receptors Channels 8:343–352

Drews J (2000) Drug discovery: a historical perspective. Science 287:1960–1964

Durocher Y, Perret S, Thibaudeau E, Gaumond MH, Kamen A, Stocco R, Abramovitz M (2000) A reporter gene assay for high-throughput screening of G-protein-coupled receptors stably or transiently expressed in HEK293 EBNA cells grown in suspension culture. Anal Biochem 284:316–326

Eglen RM (2005) Functional G protein-coupled receptor assays for primary and secondary screening. Comb Chem High Throughput Screen 8:311–318

Elagoz A, Henderson D, Babu PS, Salter S, Grahames C, Bowers L, Roy MO, Laplante P, Grazzini E, Ahmad S, Lembo PM (2004) A truncated form of CKbeta8-1 is a potent agonist for human formyl peptide-receptor-like 1 receptor. Br J Pharmacol 141:37–46

Elshourbagy NA, Ames RS, Fitzgerald LR, Foley JJ, Chambers JK, Szekeres PG, Evans NA, Schmidt DB, Buckley PT, Dytko GM, Murdock PR, Milligan G, Groarke DA, Tan KB, Shabon U, Nuthulaganti P, Wang DY, Wilson S, Bergsma DJ, Sarau HM (2000) Receptor for the pain modulatory neuropeptides FF and AF is an orphan G protein-coupled receptor. J Biol Chem 275:25965–25971

Erickson JR, Wu JJ, Goddard JG, Tigyi G, Kawanishi K, Tomei LD, Kiefer MC (1998) Edg-2/Vzg-1 couples to the yeast pheromone response pathway selectively in response to lysophosphatidic acid. J Biol Chem 273:1506–1510

Ernst S, Lange C, Wilbers A, Goebeler V, Gerke V, Rescher U (2004) An annexin 1 N-terminal peptide activates leukocytes by triggering different members of the formyl peptide receptor family. J Immunol 172:7669–7676

Evans NA, Groarke DA, Warrack J, Greenwood CJ, Dodgson K, Milligan G, Wilson S (2001) Visualizing differences in ligand-induced beta-arrestin-GFP interactions and trafficking between three recently characterized G protein-coupled receptors. J Neurochem 77:476–485

Fathi Z, Corjay MH, Shapira H, Wada E, Benya R, Jensen R, Viallet J, Sausville EA, Battey JF (1993) BRS-3: a novel bombesin receptor subtype selectively expressed in testis and lung carcinoma cells. J Biol Chem 268:5979–5984

Feighner SD, Tan CP, McKee KK, Palyha OC, Hreniuk DL, Pong SS, Austin CP, Figueroa D, MacNeil D, Cascieri MA, Nargund R, Bakshi R, Abramovitz M, Stocco R, Kargman S, O'Neill G, Van der Ploeg LH, Evans J, Patchett AA, Smith RG, Howard AD (1999) Receptor for motilin identified in the human gastrointestinal system. Science 284:2184–2188

Fiore S, Maddox JF, Perez HD, Serhan CN (1994) Identification of a human CDNA encoding a functional high affinity lipoxin A4 receptor. J Exp Med 180:253–260

Fujii R, Yoshida H, Fukusumi S, Habata Y, Hosoya M, Kawamata Y, Yano T, Hinuma S, Kitada C, Asami T, Mori M, Fujisawa Y, Fujino M (2002) Identification of a neuropeptide modified with bromine as an endogenous ligand for GPR7. J Biol Chem 277:34010–34016

Fukusumi S, Yoshida H, Fujii R, Maruyama M, Komatsu H, Habata Y, Shintani Y, Hinuma S, Fujino M (2003) A new peptidic ligand and its receptor regulating adrenal function in rats. J Biol Chem 278:46387–46395

Gabarin N, Gavish H, Muhlrad A, Chen YC, Namdar-Attar M, Nissenson RA, Chorev M, Bab I (2001) Mitogenic G(i) protein-MAP kinase signaling cascade in MC3T3-E1 osteogenic cells: activation by C-terminal pentapeptide of osteogenic growth peptide [OGP(10–14)] and attenuation of activation by CAMP. J Cell Biochem 81:594–603

Galindo E, Mendez M, Calvo S, Gonzalez-Garcia C, Cena V, Hubert P, Bader MF, Aunis D (1992) Chromostatin receptors control calcium channel activity in adrenal chromaffin cells. J Biol Chem 267:407–412

George SR, Fan T, Xie Z, Tse R, Tam V, Varghese G, O'Dowd BF (2000) Oligomerization of mu- and delta-opioid receptors. Generation of novel functional properties. J Biol Chem 275:26128–26135

Gerard NP, Lu B, Liu P, Craig S, Fujiwara Y, Okinaga S, Gerard C (2005) An anti-inflammatory function for the complement anaphylatoxin C5a-binding protein, C5L2. J Biol Chem 280:39677–39680

Goetz AS, Liacos J, Yingling J, Ignar DM (1999) A combination assay for simultaneous assessment of multiple signaling pathways. J Pharmacol Toxicol Methods 42:225–235

Gomes I, Jordan BA, Gupta A, Trapaidze N, Nagy V, Devi LA (2000) Heterodimerization of mu and delta opioid receptors: a role in opiate synergy. J Neurosci 20:RC110

Gomes I, Gupta A, Filipovska J, Szeto HH, Pintar JE, Devi LA (2004) A role for heterodimerization of mu and delta opiate receptors in enhancing morphine analgesia. Proc Natl Acad Sci USA 101:5135–5139

Gonzalez-Yanes C, Santos-Alvarez J, Sanchez-Margalet V (1999) Characterization of pancreastatin receptors and signaling in adipocyte membranes. Biochim Biophys Acta 1451:153–162

Gonzalez-Yanes C, Santos-Alvarez J, Sanchez-Margalet V (2001) Pancreastatin, a chromogranin A-derived peptide, activates Galpha(16) and phospholipase C-beta(2) by interacting with specific receptors in rat heart membranes. Cell Signal 13:43–49

Goumon Y, Angelone T, Schoentgen F, Chasserot-Golaz S, Almas B, Fukami MM, Langley K, Welters ID, Tota B, Aunis D, Metz-Boutigue MH (2004) The hippocampal cholinergic neurostimulating peptide, the N-terminal fragment of the secreted phosphatidylethanolamine-binding protein, possesses a new biological activity on cardiac physiology. J Biol Chem 279:13054–13064

Graminski GF, Jayawickreme CK, Potenza MN, Lerner MR (1993) Pigment dispersion in frog melanophores can be induced by a phorbol ester or stimulation of a recombinant receptor that activates phospholipase C. J Biol Chem 268:5957–5964

Grazzini E, Puma C, Roy MO, Yu XH, O'Donnell D, Schmidt R, Dautrey S, Ducharme J, Perkins M, Panetta R, Laird JM, Ahmad S, Lembo PM (2004) Sensory neuron-specific receptor activation elicits central and peripheral nociceptive effects in rats. Proc Natl Acad Sci USA 101:7175–7180

Green A, Milligan G, Dobias SB (1992) Gi down-regulation as a mechanism for heterologous desensitization in adipocytes. J Biol Chem 267:3223–3229

Habata Y, Fujii R, Hosoya M, Fukusumi S, Kawamata Y, Hinuma S, Kitada C, Nishizawa N, Murosaki S, Kurokawa T, Onda H, Tatemoto K, Fujino M (1999) Apelin, the natural ligand of the orphan receptor APJ, is abundantly secreted in the colostrum. Biochim Biophys Acta 1452:25–35

Hamdan FF, Audet M, Garneau P, Pelletier J, Bouvier M (2005) High-throughput screening of G protein-coupled receptor antagonists using a bioluminescence resonance energy transfer 1-based beta-arrestin2 recruitment assay. J Biomol Screen 10:463–475

Han SK, Dong X, Hwang JI, Zylka MJ, Anderson DJ, Simon MI (2002) Orphan G protein-coupled receptors MrgA1 and MrgC11 are distinctively activated by RF-amide-related peptides through the Galpha Q/11 pathway. Proc Natl Acad Sci USA 99:14740–14745

Harada M, Habata Y, Hosoya M, Nishi K, Fujii R, Kobayashi M, Hinuma S (2004) N-formylated humanin activates both formyl peptide receptor-like 1 and 2. Biochem Biophys Res Commun 324:255–261

He W, Miao FJ, Lin DC, Schwandner RT, Wang Z, Gao J, Chen JL, Tian H, Ling L (2004) Citric acid cycle intermediates as ligands for orphan G-protein-coupled receptors. Nature 429:188–193

Heise CE, O'Dowd BF, Figueroa DJ, Sawyer N, Nguyen T, Im DS, Stocco R, Bellefeuille JN, Abramovitz M, Cheng R, Williams DL Jr, Zeng Z, Liu Q, Ma L, Clements MK, Coulombe N, Liu Y, Austin CP, George SR, O'Neill GP, Metters KM, Lynch KR, Evans JF (2000) Characterization of the human cysteinyl leukotriene 2 receptor. J Biol Chem 275:30531–30536

Hemmila II (1999) LANCEtrade mark: homogeneous assay platform for HTS. J Biomol Screen 4:303–308

Heusler P, Pauwels PJ, Wurch T, Newman-Tancredi A, Tytgat J, Colpaert FC, Cussac D (2005) Differential ion current activation by human 5-HT(1A) receptors in Xenopus Oocytes: evidence for agonist-directed trafficking of receptor signalling. Neuropharmacology 49:963–976

Hill J, Duckworth M, Murdock P, Rennie G, Sabido-David C, Ames RS, Szekeres P, Wilson S, Bergsma DJ, Gloger IS, Levy DS, Chambers JK, Muir AI (2001) Molecular cloning and functional characterization of MCH2, a novel human MCH receptor. J Biol Chem 276:20125–20129

Hinuma S, Habata Y, Fujii R, Kawamata Y, Hosoya M, Fukusumi S, Kitada C, Masuo Y, Asano T, Matsumoto H, Sekiguchi M, Kurokawa T, Nishimura O, Onda H, Fujino M (1998) A prolactin-releasing peptide in the brain. Nature 393:272–276

Hirai H, Tanaka K, Yoshie O, Ogawa K, Kenmotsu K, Takamori Y, Ichimasa M, Sugamura K, Nakamura M, Takano S, Nagata K (2001) Prostaglandin D2 selectively induces chemotaxis in T helper type 2 cells, eosinophils, and basophils via seven-transmembrane receptor CRTH2. J Exp Med 193:255–261

Hiraiwa M, Campana WM, Martin BM, O'Brien JS (1997) Prosaposin receptor: evidence for a G-protein-associated receptor. Biochem Biophys Res Commun 240:415–418

Hirasawa A, Tsumaya K, Awaji T, Katsuma S, Adachi T, Yamada M, Sugimoto Y, Miyazaki S, Tsujimoto G (2005) Free fatty acids regulate gut incretin glucagon-like peptide-1 secretion through GPR120. Nat Med 11:90–94

Hollopeter G, Jantzen HM, Vincent D, Li G, England L, Ramakrishnan V, Yang RB, Nurden P, Nurden A, Julius D, Conley PB (2001) Identification of the platelet ADP receptor targeted by antithrombotic drugs. Nature 409:202–207

Hosoi T, Koguchi Y, Sugikawa E, Chikada A, Ogawa K, Tsuda N, Suto N, Tsunoda S, Taniguchi T, Ohnuki T (2002) Identification of a novel human eicosanoid receptor coupled to G(i/o). J Biol Chem 277:31459–31465

Howard AD, Wang R, Pong SS, Mellin TN, Strack A, Guan XM, Zeng Z, Williams DL Jr, Feighner SD, Nunes CN, Murphy B, Stair JN, Yu H, Jiang Q, Clements MK, Tan CP, McKee KK, Hreniuk DL, McDonald TP, Lynch KR, Evans JF, Austin CP, Caskey CT, Van der Ploeg LH, Liu Q (2000) Identification of receptors for neuromedin U and its role in feeding. Nature 406:70–74

Hsu CS, Ho YC, Wang KC, Hu YC (2004) Investigation of optimal transduction conditions for baculovirus-mediated gene delivery into mammalian cells. Biotechnol Bioeng 88:42–51

Hsu SY, Nakabayashi K, Nishi S, Kumagai J, Kudo M, Sherwood OD, Hsueh AJ (2002) Activation of orphan receptors by the hormone relaxin. Science 295:671–674

Idzko M, Panther E, Bremer HC, Windisch W, Sorichter S, Herouy Y, Elsner P, Mockenhaupt M, Girolomoni G, Norgauer J (2004) Inosine stimulates chemotaxis, Ca2+-transients and actin polymerization in immature human dendritic cells via a pertussis toxin-sensitive mechanism independent of adenosine receptors. J Cell Physiol 199:149–156

Im DS, Heise CE, Ancellin N, O'Dowd BF, Shei GJ, Heavens RP, Rigby MR, Hla T, Mandala S, McAllister G, George SR, Lynch KR (2000a) Characterization of a novel sphingosine 1-phosphate receptor, Edg-8. J Biol Chem 275:14281–14286

Im DS, Heise CE, Harding MA, George SR, O'Dowd BF, Theodorescu DD, Lynch KR (2000b) Molecular cloning and characterization of a lysophosphatidic acid receptor, Edg-7, expressed in prostate. Mol Pharmacol 57:753–759

Im DS, Heise CE, Nguyen T, O'Dowd BF, Lynch KR (2001) Identification of a molecular target of psychosine and its role in globoid cell formation. J Cell Biol 153:429–434

Inbe H, Watanabe S, Miyawaki M, Tanabe E, Encinas JA (2004) Identification and characterization of a cell-surface receptor, P2Y15, for AMP and adenosine. J Biol Chem 279:19790–19799

Ishii S, Kihara Y, Shimizu T (2005) Identification of T cell death-associated gene 8 (TDAG8) as a novel acid sensing G-protein-coupled receptor. J Biol Chem 280:9083–9087

Itoh Y, Kawamata Y, Harada M, Kobayashi M, Fujii R, Fukusumi S, Ogi K, Hosoya M, Tanaka Y, Uejima H, Tanaka H, Maruyama M, Satoh R, Okubo S, Kizawa H, Komatsu H, Matsumura F, Noguchi Y, Shinohara T, Hinuma S, Fujisawa Y, Fujino M (2003) Free fatty acids regulate insulin secretion from pancreatic beta cells through GPR40. Nature 422:173–176

Jayawickreme CK, Sauls H, Bolio N, Ruan J, Moyer M, Burkhart W, Marron B, Rimele T, Shaffer J (1999) Use of a cell-based, lawn format assay to rapidly screen a 442,368 bead-based peptide library. J Pharmacol Toxicol Methods 42:189–197

Johnson EC, Bohn LM, Barak LS, Birse RT, Nassel DR, Caron MG, Taghert PH (2003) Identification of Drosophila neuropeptide receptors by G protein-coupled receptors-beta-arrestin2 interactions. J Biol Chem 278:52172–52178

Jones CE, Holden S, Tenaillon L, Bhatia U, Seuwen K, Tranter P, Turner J, Kettle R, Bouhelal R, Charlton S, Nirmala NR, Jarai G, Finan P (2003) Expression and characterization of a 5-Oxo-6E,8Z,11Z,14Z-eicosatetraenoic acid receptor highly expressed on human eosinophils and neutrophils. Mol Pharmacol 63:471–477

Jones KA, Borowsky B, Tamm JA, Craig DA, Durkin MM, Dai M, Yao WJ, Johnson M, Gunwaldsen C, Huang LY, Tang C, Shen Q, Salon JA, Morse K, Laz T, Smith KE, Nagarathnam D, Noble SA, Branchek TA, Gerald C (1998) GABA(B) receptors function as a heteromeric assembly of the subunits GABA(B)R1 and GABA(B)R2. Nature 396:674–679

Jonsson JR, Moschen AR, Hickman IJ, Richardson MM, Kaser S, Clouston AD, Powell EE, Tilg H (2005) Adiponectin and its receptors in patients with chronic hepatitis C. J Hepatol 43:929–936

Jordan BA, Devi LA (1999) G-protein-coupled receptor heterodimerization modulates receptor function. Nature 399:697–700

Kabarowski JH, Zhu K, Le LQ, Witte ON, Xu Y (2001) Lysophosphatidylcholine as a ligand for the immunoregulatory receptor G2A. Science 293:702–705

Kalant D, Cain SA, Maslowska M, Sniderman AD, Cianflone K, Monk PN (2003) The chemoattractant receptor-like protein C5L2 binds the C3a Des-Arg77/acylation-stimulating protein. J Biol Chem 278:11123–11129

Kalant D, MacLaren R, Cui W, Samanta R, Monk PN, Laporte SA, Cianflone K (2005) C5L2 Is a functional receptor for acylation-stimulating protein. J Biol Chem 280:23936–23944

Kamohara M, Takasaki J, Matsumoto M, Saito T, Ohishi T, Ishii H, Furuichi K (2000) Molecular cloning and characterization of another leukotriene B4 receptor. J Biol Chem 275:27000–27004

Kamohara M, Matsuo A, Takasaki J, Kohda M, Matsumoto M, Matsumoto S, Soga T, Hiyama H, Kobori M, Katou M (2005) Identification of MrgX2 as a human G-protein-coupled receptor for proadrenomedullin N-terminal peptides. Biochem Biophys Res Commun 330:1146–1152

Kanda N, Watanabe S (2003) 17beta-estradiol enhances the production of nerve growth factor in THP-1-derived macrophages or peripheral blood monocyte-derived macrophages. J Invest Dermatol 121:771–780

Kaupmann K, Malitschek B, Schuler V, Heid J, Froestl W, Beck P, Mosbacher J, Bischoff S, Kulik A, Shigemoto R, Karschin A, Bettler B (1998) GABA(B)-receptor subtypes assemble into functional heteromeric complexes. Nature 396:683–687

Kawamata Y, Fujii R, Hosoya M, Harada M, Yoshida H, Miwa M, Fukusumi S, Habata Y, Itoh T, Shintani Y, Hinuma S, Fujisawa Y, Fujino M (2003) A G protein-coupled receptor responsive to bile acids. J Biol Chem 278:9435–9440

Kojima M, Hosoda H, Date Y, Nakazato M, Matsuo H, Kangawa K (1999) Ghrelin is a growth-hormone-releasing acylated peptide from stomach. Nature 402:656–660

Kojima M, Haruno R, Nakazato M, Date Y, Murakami N, Hanada R, Matsuo H, Kangawa K (2000) Purification and identification of neuromedin U as an endogenous ligand for an orphan receptor GPR66 (FM3). Biochem Biophys Res Commun 276:435–438

Kost TA, Condreay JP, Jarvis DL (2005) Baculovirus as versatile vectors for protein expression in insect and mammalian cells. Nat Biotechnol 23:567–575

Kostenis E (2001) Is Galpha16 the optimal tool for fishing ligands of orphan G-protein-coupled receptors? Trends Pharmacol Sci 22:560–564

Kotarsky K, Nilsson NE, Flodgren E, Owman C, Olde B (2003) A human cell surface receptor activated by free fatty acids and thiazolidinedione drugs. Biochem Biophys Res Commun 301:406–410

Kraakman L, Lemaire K, Ma P, Teunissen AW, Donaton MC, Van Dijck P, Winderickx J, de Winde JH, Thevelein JM (1999) A *Saccharomyces cerevisiae* G-protein coupled receptor, Gpr1, is specifically required for glucose activation of the CAMP pathway during the transition to growth on glucose. Mol Microbiol 32:1002–1012

Krasnoperov VG, Bittner MA, Beavis R, Kuang Y, Salnikow KV, Chepurny OG, Little AR, Plotnikov AN, Wu D, Holz RW, Petrenko AG (1997) Alpha-latrotoxin stimulates exocytosis by the interaction with a neuronal G-protein-coupled receptor. Neuron 18:925–937

Kunapuli P, Ransom R, Murphy KL, Pettibone D, Kerby J, Grimwood S, Zuck P, Hodder P, Lacson R, Hoffman I, Inglese J, Strulovici B (2003) Development of an intact cell reporter gene beta-lactamase assay for G protein-coupled receptors for high-throughput screening. Anal Biochem 314:16–29

Lander ES, Linton LM, Birren B et al. (2001) Initial sequencing and analysis of the human genome. Nature 409:860–921

Lee MJ, Van Brocklyn JR, Thangada S, Liu CH, Hand AR, Menzeleev R, Spiegel S, Hla T (1998) Sphingosine-1-phosphate as a ligand for the g protein-coupled receptor EDG-1. Science 279:1552–1555

Lee SP, Xie Z, Varghese G, Nguyen T, O'Dowd BF, George SR (2000) Oligomerization of dopamine and serotonin receptors. Neuropsychopharmacology 23:S32–S40

Lembo PM, Grazzini E, Cao J, Hubatsch DA, Pelletier M, Hoffert C, St Onge S, Pou C, Labrecque J, Groblewski T, O'Donnell D, Payza K, Ahmad S, Walker P (1999) The receptor for the orexigenic peptide melanin-concentrating hormone is a G-protein-coupled receptor. Nat Cell Biol 1:267–271

Lembo PM, Grazzini E, Groblewski T, O'Donnell D, Roy MO, Zhang J, Hoffert C, Cao J, Schmidt R, Pelletier M, Labarre M, Gosselin M, Fortin Y, Banville D, Shen SH, Strom P, Payza K, Dray A, Walker P, Ahmad S (2002) Proenkephalin A gene products activate a new family of sensory neuron-specific GPCRs. Nat Neurosci 5:201–209

Lerner MR (1994) Tools for investigating functional interactions between ligands and G-protein-coupled receptors. Trends Neurosci 17:142–146

Liu C, Ma X, Jiang X, Wilson SJ, Hofstra CL, Blevitt J, Pyati J, Li X, Chai W, Carruthers N, Lovenberg TW (2001) Cloning and pharmacological characterization of a fourth histamine receptor (H(4)) expressed in bone marrow. Mol Pharmacol 59:420–426

Liu PS, Wang PY (2004) DHEA attenuates catecholamine secretion from bovine adrenal chromaffin cells. J Biomed Sci 11:200–205

Lorenzen A, Stannek C, Lang H, Andrianov V, Kalvinsh I, Schwabe U (2001) Characterization of a G protein-coupled receptor for nicotinic acid. Mol Pharmacol 59:349–357

Luo XH, Guo LJ, Yuan LQ, Xie H, Zhou HD, Wu XP, Liao EY (2005) Adiponectin stimulates human osteoblasts proliferation and differentiation via the MAPK signaling pathway. Exp Cell Res 309:99–109

Lynch KR, O'Neill GP, Liu Q, Im DS, Sawyer N, Metters KM, Coulombe N, Abramovitz M, Figueroa DJ, Zeng Z, Connolly BM, Bai C, Austin CP, Chateauneuf A, Stocco R, Greig GM, Kargman S, Hooks SB, Hosfield E, Williams DL Jr, Ford-Hutchinson AW, Caskey CT, Evans JF (1999) Characterization of the human cysteinyl leukotriene CysLT1 receptor. Nature 399:789–793

Maeda N, Shimomura I, Kishida K, Nishizawa H, Matsuda M, Nagaretani H, Furuyama N, Kondo H, Takahashi M, Arita Y, Komuro R, Ouchi N, Kihara S, Tochino Y, Okutomi K, Horie M, Takeda S, Aoyama T, Funahashi T, Matsuzawa Y (2002) Diet-induced insulin resistance in mice lacking adiponectin/ACRP30. Nat Med 8:731–737

Maggiolini M, Vivacqua A, Fasanella G, Recchia AG, Sisci D, Pezzi V, Montanaro D, Musti AM, Picard D, Ando S (2004) The G protein-coupled receptor GPR30 mediates C-Fos up-regulation by 17beta-estradiol and phytoestrogens in breast cancer cells. J Biol Chem 279:27008–27016

Maidan MM, De Rop L, Serneels J, Exler S, Ru S, Tournu H, Thevelein JM, Van Dijck P (2005) The G protein-coupled receptor Gpr1 and the Galpha protein Gpa2 Act through the CAMP-protein kinase A pathway to induce morphogenesis in *Candida albicans*. Mol Biol Cell 16:1971–1986

Maruyama T, Miyamoto Y, Nakamura T, Tamai Y, Okada H, Sugiyama E, Nakamura T, Itadani H, Tanaka K (2002) Identification of membrane-type receptor for bile acids (M-BAR). Biochem Biophys Res Commun 298:714–719

Maslowska M, Legakis H, Assadi F, Cianflone K (2006) Targeting the signaling pathway of acylation stimulating protein. J Lipid Res 47:643–652

McClintock TS, Graminski GF, Potenza MN, Jayawickreme CK, Roby-Shemkovitz A, Lerner MR (1993) Functional expression of recombinant G-protein-coupled receptors monitored by video imaging of pigment movement in melanophores. Anal Biochem 209:298–305

McLatchie LM, Fraser NJ, Main MJ, Wise A, Brown J, Thompson N, Solari R, Lee MG, Foord SM (1998) RAMPs regulate the transport and ligand specificity of the calcitonin-receptor-like receptor. Nature 393:333–339

Meder W, Wendland M, Busmann A, Kutzleb C, Spodsberg N, John H, Richter R, Schleuder D, Meyer M, Forssmann WG (2003) Characterization of human circulating TIG2 as a ligand for the orphan receptor ChemR23. FEBS Lett 555:495–499

Medici R, Bianchi E, Di Segni G, Tocchini-Valentini GP (1997) Efficient signal transduction by a chimeric yeast-mammalian G protein alpha subunit Gpa1-Gsalpha covalently fused to the yeast receptor Ste2. EMBO J 16:7241–7249

Meunier JC, Mollereau C, Toll L, Suaudeau C, Moisand C, Alvinerie P, Butour JL, Guillemot JC, Ferrara P, Monsarrat B et al (1995) Isolation and structure of the endogenous agonist of opioid receptor-like ORL1 receptor. Nature 377:532–535

Milasta S, Pediani J, Appelbe S, Trim S, Wyatt M, Cox P, Fidock M, Milligan G (2006) Interactions between the Mas-related receptors MrgD and MrgE alter signalling and trafficking of MrgD. Mol Pharmacol 69:479–491

Milligan G (2000) Insights into ligand pharmacology using receptor-G-protein fusion proteins. Trends Pharmacol Sci 21:24–28

Milligan G (2003) Principles: extending the utility of [35S]GTP gamma S binding assays. Trends Pharmacol Sci 24:87–90

Mody SM, Ho MK, Joshi SA, Wong YH (2000) Incorporation of Galpha(z)-specific sequence at the carboxyl terminus increases the promiscuity of Galpha(16) toward G(i)-coupled receptors. Mol Pharmacol 57:13–23

Mori M, Sugo T, Abe M, Shimomura Y, Kurihara M, Kitada C, Kikuchi K, Shintani Y, Kurokawa T, Onda H, Nishimura O, Fujino M (1999) Urotensin II is the endogenous ligand of a G-protein-coupled orphan receptor, SENR (GPR14). Biochem Biophys Res Commun 265:123–129

Morse KL, Behan J, Laz TM, West RE Jr, Greenfeder SA, Anthes JC, Umland S, Wan Y, Hipkin RW, Gonsiorek W, Shin N, Gustafson EL, Qiao X, Wang S, Hedrick JA, Greene J, Bayne M, Monsma FJ Jr (2001) Cloning and characterization of a novel human histamine receptor. J Pharmacol Exp Ther 296:1058–1066

Nakamura T, Itadani H, Hidaka Y, Ohta M, Tanaka K (2000) Molecular cloning and characterization of a new human histamine receptor, HH4R. Biochem Biophys Res Commun 279:615–620

Nguyen T, Shapiro DA, George SR, Setola V, Lee DK, Cheng R, Rauser L, Lee SP, Lynch KR, Roth BL, O'Dowd BF (2001) Discovery of a novel member of the histamine receptor family. Mol Pharmacol 59:427–433

Nothacker HP, Wang Z, McNeill AM, Saito Y, Merten S, O'Dowd B, Duckles SP, Civelli O (1999) Identification of the natural ligand of an orphan G-protein-coupled receptor involved in the regulation of vasoconstriction. Nat Cell Biol 1:383–385

Nothacker HP, Wang Z, Zhu Y, Reinscheid RK, Lin SH, Civelli O (2000) Molecular cloning and characterization of a second human cysteinyl leukotriene receptor: discovery of a subtype selective agonist. Mol Pharmacol 58:1601–1608

O'Dowd BF, Ji X, Alijaniaram M, Rajaram RD, Kong MM, Rashid A, Nguyen T, George SR (2005) Dopamine receptor oligomerization visualized in living cells. J Biol Chem 280:37225–37235

Oakley RH, Hudson CC, Cruickshank RD, Meyers DM, Payne RE Jr, Rhem SM, Loomis CR (2002) The cellular distribution of fluorescently labeled arrestins provides a robust, sensitive, and universal assay for screening g protein-coupled receptors. Assay Drug Dev Technol 1:21–30

Oda T, Morikawa N, Saito Y, Masuho Y, Matsumoto S (2000) Molecular cloning and characterization of a novel type of histamine receptor preferentially expressed in leukocytes. J Biol Chem 275:36781–36786

Offermanns S, Simon MI (1995) G alpha 15 and G alpha 16 couple a wide variety of receptors to phospholipase C. J Biol Chem 270:15175–15180

Ohtaki T, Shintani Y, Honda S, Matsumoto H, Hori A, Kanehashi K, Terao Y, Kumano S, Takatsu Y, Masuda Y, Ishibashi Y, Watanabe T, Asada M, Yamada T, Suenaga M, Kitada C, Usuki S, Kurokawa T, Onda H, Nishimura O, Fujino M (2001) Metastasis suppressor gene KiSS-1 encodes peptide ligand of a G-protein-coupled receptor. Nature 411:613–617

Okinaga S, Slattery D, Humbles A, Zsengeller Z, Morteau O, Kinrade MB, Brodbeck RM, Krause JE, Choe HR, Gerard NP, Gerard C (2003) C5L2, a nonsignaling C5A binding protein. Biochemistry 42:9406–9415

Overton HA, Babbs AJ, Doel SM, Fyfe MC, Gardner LS, Griffin G, Jackson HC, Procter MJ, Rasamison CM, Tang-Christensen M, Widdowson PS, Williams GM, Reynet C (2006) Deorphanization of a G protein-coupled receptor for oleoylethanolamide and its use in the discovery of small-molecule hypophagic agents. Cell Metab 3:167–175

Pausch MH, Lai M, Tseng E, Paulsen J, Bates B, Kwak S (2004) Functional expression of human and mouse P2Y12 receptors in *Saccharomyces cerevisiae*. Biochem Biophys Res Commun 324:171–177

Perroy J, Adam L, Qanbar R, Chenier S, Bouvier M (2003) Phosphorylation-independent desensitization of GABA(B) Receptor by GRK4. EMBO J 22:3816–3824

Pike NB, Wise A (2004) Identification of a nicotinic acid receptor: is this the molecular target for the oldest lipid-lowering drug? Curr Opin Investig Drugs 5:271–275

Rees S, Martin DP, Scott SV, Brown SH, Fraser N, O'Shaughnessy C, Beresford IJ (2001) Development of a homogeneous MAp kinase reporter gene screen for the identification of agonists and antagonists at the CXCR1 chemokine receptor. J Biomol Screen 6:19–27

Rezgaoui M, Susens U, Ignatov A, Gelderblom M, Glassmeier G, Franke I, Urny J, Imai Y, Takahashi R and Schaller HC (2006) The neuropeptide head activator is a high-affinity ligand for the orphan G-Protein-coupled receptor GPR37. J Cell Sci 119:542–549

Robas N, Mead E and Fidock M (2003) MrgX2 is a high potency cortistatin receptor expressed in dorsal root ganglion. J Biol Chem 278:44400–44404

Rocheville M, Lange DC, Kumar U, Patel SC, Patel RC, Patel YC (2000) Receptors for dopamine and somatostatin: formation of hetero-oligomers with enhanced functional activity. Science 288:154–157

Sakurai T, Amemiya A, Ishii M, Matsuzaki I, Chemelli RM, Tanaka H, Williams SC, Richardson JA, Kozlowski GP, Wilson S, Arch JR, Buckingham RE, Haynes AC, Carr SA, Annan RS, McNulty DE, Liu WS, Terrett JA, Elshourbagy NA, Bergsma DJ, Yanagisawa M (1998) Orexins and orexin receptors: a family of hypothalamic neuropeptides and G protein-coupled receptors that regulate feeding behavior. Cell 92:573–585

Santos-Alvarez J, Gonzalez-Yanes C, Sanchez-Margalet V (1998) Pancreastatin receptor is coupled to a guanosine triphosphate-binding protein of the G(q/11)alpha family in rat liver membranes. Hepatology 27:608–614

Satoh F, Takahashi K, Murakami O, Totsune K, Ohneda M, Mizuno Y, Sone M, Miura Y, Takase S, Hayashi Y, Sasano H, Mouri T (1997) Cerebellin and cerebellin MRNA in the human brain, adrenal glands and the tumour tissues of adrenal tumour, ganglioneuroblastoma and neuroblastoma. J Endocrinol 154:27–34

Schrick K, Garvik B, Hartwell LH (1997) Mating in *Saccharomyces cerevisiae*: the role of the pheromone signal transduction pathway in the chemotropic response to pheromone. Genetics 147:19–32

Shenoy SK, Lefkowitz RJ (2005) Seven-transmembrane receptor signaling through beta-arrestin. Sci STKE 2005:cm10

Shenoy SK, Drake MT, Nelson CD, Houtz DA, Xiao K, Madabushi S, Reiter E, Premont RT, Lichtarge O, Lefkowitz RJ (2006) Beta-arrestin-dependent, G protein-independent ERK1/2 activation by the beta2 adrenergic receptor. J Biol Chem 281:1261–1273

Shichiri M, Ishimaru S, Ota T, Nishikawa T, Isogai T and Hirata Y (2003) Salusins: newly identified bioactive peptides with hemodynamic and mitogenic activities. Nat Med 9:1166–1172

Shimomura Y, Harada M, Goto M, Sugo T, Matsumoto Y, Abe M, Watanabe T, Asami T, Kitada C, Mori M, Onda H, Fujino M (2002) Identification of neuropeptide W as the endogenous ligand for orphan G-protein-coupled receptors GPR7 and GPR8. J Biol Chem 277:35826–35832

Shinohara T, Harada M, Ogi K, Maruyama M, Fujii R, Tanaka H, Fukusumi S, Komatsu H, Hosoya M, Noguchi Y, Watanabe T, Moriya T, Itoh Y, Hinuma S (2004) Identification of a G protein-coupled receptor specifically responsive to beta-alanine. J Biol Chem 279:23559–23564

Simmons MA (2005) Functional selectivity, ligand-directed trafficking, conformation-specific agonism: what's in a name? Mol Interv 5:154–157

Smit MJ, Verzijl D, Casarosa P, Navis M, Timmerman H, Leurs R (2002) Kaposi's sarcoma-associated herpesvirus-encoded G protein-coupled receptor ORF74 constitutively activates P44/P42 MAPK and Akt Via G(i) and phospholipase C-dependent signaling pathways. J Virol 76:1744–1752

Soga T, Kamohara M, Takasaki J, Matsumoto S, Saito T, Ohishi T, Hiyama H, Matsuo A, Matsushime H, Furuichi K (2003) Molecular identification of nicotinic acid receptor. Biochem Biophys Res Commun 303:364–369

Stables J, Green A, Marshall F, Fraser N, Knight E, Sautel M, Milligan G, Lee M, Rees S (1997) A bioluminescent assay for agonist activity at potentially any G-protein-coupled receptor. Anal Biochem 252:115–126

Su SB, Gong W, Gao JL, Shen W, Murphy PM, Oppenheim JJ, Wang JM (1999) A seven-transmembrane, G protein-coupled receptor, FPRL1, mediates the chemotactic activity of serum amyloid a for human phagocytic cells. J Exp Med 189:395–402

Szekeres PG, Muir AI, Spinage LD, Miller JE, Butler SI, Smith A, Rennie GI, Murdock PR, Fitzgerald LR, Wu H, McMillan LJ, Guerrera S, Vawter L, Elshourbagy NA, Mooney JL, Bergsma DJ, Wilson S, Chambers JK (2000) Neuromedin U is a potent agonist at the orphan G protein-coupled receptor FM3. J Biol Chem 275:20247–20250

Takeda S, Yamamoto A, Okada T, Matsumura E, Nose E, Kogure K, Kojima S, Haga T (2003) Identification of surrogate ligands for orphan G protein-coupled receptors. Life Sci 74:367–377

Tanaka H, Yoshida T, Miyamoto N, Motoike T, Kurosu H, Shibata K, Yamanaka A, Williams SC, Richardson JA, Tsujino N, Garry MG, Lerner MR, King DS, O'Dowd BF, Sakurai T, Yanagisawa M (2003) Characterization of a family of endogenous neuropeptide ligands for the G protein-coupled receptors GPR7 and GPR8. Proc Natl Acad Sci USA 100:6251–6256

Tatemoto K, Hosoya M, Habata Y, Fujii R, Kakegawa T, Zou MX, Kawamata Y, Fukusumi S, Hinuma S, Kitada C, Kurokawa T, Onda H, Fujino M (1998) Isolation and characterization of a novel endogenous peptide ligand for the human APJ receptor. Biochem Biophys Res Commun 251:471–476

Tunaru S, Kero J, Schaub A, Wufka C, Blaukat A, Pfeffer K, Offermanns S (2003) PUMA-G and HM74 are receptors for nicotinic acid and mediate its anti-lipolytic effect. Nat Med 9:352–355

Ueda H, Inoue M (2000) In vivo signal transduction of nociceptive response by kyotorphin (tyrosine-arginine) through Galpha(i)- and inositol trisphosphate-mediated Ca(2+) Influx. Mol Pharmacol 57:108–115

Venter JC, Adams MD, Myers EW et al. (2001) The sequence of the human genome. Science 291:1304–1351

Vischer HF, Leurs R, Smit MJ (2006) HCMV-encoded G-Protein-coupled receptors as constitutively active modulators of cellular signaling networks. Trends Pharmacol Sci 27:56–63

Vrecl M, Jorgensen R, Pogacnik A, Heding A (2004) Development of a BRET2 screening assay using beta-arrestin 2 mutants. J Biomol Screen 9:322–333

Wellendorph P, Hansen KB, Balsgaard A, Greenwood JR, Egebjerg J, Brauner-Osborne H (2005) Deorphanization of GPRC6A: a promiscuous L-alpha-amino acid receptor with preference for basic amino acids. Mol Pharmacol 67:589–597

White JH, Wise A, Main MJ, Green A, Fraser NJ, Disney GH, Barnes AA, Emson P, Foord SM, Marshall FH (1998) Heterodimerization is required for the formation of a functional GABA(B) receptor. Nature 396:679–682

Whitman SC, Daugherty A, Post SR (2000) Macrophage colony-stimulating factor rapidly enhances beta-migrating very low density lipoprotein metabolism in macrophages through activation of a Gi/o protein signaling pathway. J Biol Chem 275:35807–35813

Windh RT, Lee MJ, Hla T, An S, Barr AJ, Manning DR (1999) Differential coupling of the sphingosine 1-phosphate receptors Edg-1, Edg-3, and H218/Edg-5 to the G(i), G(q), and G(12) families of heterotrimeric G proteins. J Biol Chem 274:27351–27358

Wise A, Gearing K, Rees S (2002) Target validation of G-protein coupled receptors. Drug Discov Today 7:235–246

Wise A, Foord SM, Fraser NJ, Barnes AA, Elshourbagy N, Eilert M, Ignar DM, Murdock PR, Steplewski K, Green A, Brown AJ, Dowell SJ, Szekeres PG, Hassall DG, Marshall FH, Wilson S, Pike NB (2003) Molecular identification of high and low affinity receptors for nicotinic acid. J Biol Chem 278:9869–9874

Wise A, Jupe SC, Rees S (2004) The identification of ligands at orphan G-protein coupled receptors. Annu Rev Pharmacol Toxicol 44:43–66

Wittau N, Grosse R, Kalkbrenner F, Gohla A, Schultz G, Gudermann T (2000) The galanin receptor type 2 initiates multiple signaling pathways in small cell lung cancer cells by coupling to G(q), G(i) and G(12) proteins. Oncogene 19:4199–4209

Xu Y, Zhu K, Hong G, Wu W, Baudhuin LM, Xiao Y, Damron DS (2000) Sphingosylphosphorylcholine is a ligand for ovarian cancer G-protein-coupled receptor 1. Nat Cell Biol 2:261–267

Xu YL, Reinscheid RK, Huitron-Resendiz S, Clark SD, Wang Z, Lin SH, Brucher FA, Zeng J, Ly NK, Henriksen SJ, De Lecea L, Civelli O (2004) Neuropeptide S: a neuropeptide promoting arousal and anxiolytic-like effects. Neuron 43:487–497

Xue Y, Batlle M, Hirsch JP (1998) GPR1 encodes a putative G protein-coupled receptor that associates with the Gpa2p Galpha subunit and functions in a Ras-Independent pathway. EMBO J 17:1996–2007

Yamauchi T, Kamon J, Ito Y, Tsuchida A, Yokomizo T, Kita S, Sugiyama T, Miyagishi M, Hara K, Tsunoda M, Murakami K, Ohteki T, Uchida S, Takekawa S, Waki H, Tsuno NH, Shibata Y, Terauchi Y, Froguel P, Tobe K, Koyasu S, Taira K, Kitamura T, Shimizu T, Nagai R, Kadowaki T (2003) Cloning of adiponectin receptors that mediate antidiabetic metabolic effects. Nature 423:762–769

Yamazaki Y, Kon J, Sato K, Tomura H, Sato M, Yoneya T, Okazaki H, Okajima F, Ohta H (2000) Edg-6 as a putative sphingosine 1-phosphate receptor coupling to Ca(2+) signaling pathway. Biochem Biophys Res Commun 268:583–589

Ying G, Iribarren P, Zhou Y, Gong W, Zhang N, Yu ZX, Le Y, Cui Y, Wang JM (2004) Humanin, a newly identified neuroprotective factor, uses the G protein-coupled formylpeptide receptor-like-1 as a functional receptor. J Immunol 172:7078–7085

Zeilhofer HU, Selbach UM, Guhring H, Erb K, Ahmadi S (2000) Selective suppression of inhibitory synaptic transmission by nocistatin in the rat spinal cord dorsal horn. J Neurosci 20:4922–4929

Zhang JV, Ren PG, Avsian-Kretchmer O, Luo CW, Rauch R, Klein C, Hsueh AJ (2005a) Obestatin, a peptide encoded by the ghrelin gene, opposes ghrelin's effects on food intake. Science 310:996–999

Zhang L, Taylor N, Xie Y, Ford R, Johnson J, Paulsen JE, Bates B (2005b) Cloning and expression of MRG receptors in macaque, mouse, and human. Brain Res Mol Brain Res 133:187–197

Zhu K, Baudhuin LM, Hong G, Williams FS, Cristina KL, Kabarowski JH, Witte ON, Xu Y (2001a) Sphingosylphosphorylcholine and lysophosphatidylcholine are ligands for the G protein-coupled receptor GPR4. J Biol Chem 276:41325–41335

Zhu Y, Michalovich D, Wu H, Tan KB, Dytko GM, Mannan IJ, Boyce R, Alston J, Tierney LA, Li X, Herrity NC, Vawter L, Sarau HM, Ames RS, Davenport CM, Hieble JP, Wilson S, Bergsma DJ, Fitzgerald LR (2001b) Cloning, expression, and pharmacological characterization of a novel human histamine receptor. Mol Pharmacol 59:434–441

Zhu WZ, Chakir K, Zhang S, Yang D, Lavoie C, Bouvier M, Hebert TE, Lakatta EG, Cheng H, Xiao RP (2005) Heterodimerization of beta1- and beta2-adrenergic receptor subtypes optimizes beta-adrenergic modulation of cardiac contractility. Circ Res 97:244–251

Zylka MJ, Dong X, Southwell AL, Anderson DJ (2003) Atypical expansion in mice of the sensory neuron-specific Mrg G protein-coupled receptor family. Proc Natl Acad Sci USA 100:10043–10048

Ernst Schering Foundation Symposium Proceedings, Vol. 2, pp. 145–161
DOI 10.1007/2789_2006_007

Published Online: 16 May 2007

The Role of GPCR Dimerisation/Oligomerisation in Receptor Signalling

G. Milligan(✉), M. Canals, J.D. Pediani, J. Ellis, J.F. Lopez-Gimenez

Molecular Pharmacology Group, Division of Biochemistry and Molecular Biology, Institute of Biomedical and Life Sciences, University of Glasgow, G12 8QQ Glasgow, Scotland, UK
email: *g.milligan@bio.gla.ac.uk*

Abstract. A wide range of techniques have been employed to examine the quaternary structure of G-protein-coupled receptors (GPCRs). Although it is well established that homo-dimerisation is common, recent studies have sought to explore the physical basis of these interactions and the role of dimerisation in signal transduction. Growing evidence hints at the existence of higher-order organisation of individual GPCRs and the potential for hetero-dimerisation between pairs of co-expressed GPCRs. Here we consider how both homo-dimerisation/oligomerisation and hetero-dimerisation can regulate signal transduction through GPCRs and the potential consequences of this for function of therapeutic medicines that target GPCRs. Hetero-dimerisation is not the sole means by which co-expressed GPCRs may regulate the function of one another. Heterologous desensitisation may be at least as important and we also consider if this can be the basis for physiological antagonism between pairs of co-expressed GPCRs.

Although there may be exceptions (Meyer et al. 2006), a great deal of recent evidence has indicated that most G-protein-coupled receptors (GPCRs) do not exist as monomers but rather as dimers or, potentially, within higher-order oligomers (Milligan 2004b; Park et al. 2004). Support for such models has been provided by a range of studies employing different approaches, including co-immunoprecipitation of differentially epitope-tagged but co-expressed forms of the same GPCR, co-operativity in ligand binding and a variety of resonance energy transfer techniques (Milligan and Bouvier 2005). Only for the photon receptor rhodopsin has the organisational structure of a GPCR been studied in situ. The application of atomic force microscopy to murine rod outer segment discs indicated that rhodopsin is organised in a series of parallel arrays of dimers (Liang et al. 2003) and based on this, molecular models were constructed to try to define and interpret regions of contact between the monomers (Fotiadis et al. 2004). Only for relatively few other GPCRs are details of the molecular basis of dimerisation available but within this limited data set, recent studies on the dopamine D2 receptor suggest a means by which information on the binding of an agonist can be transmitted between the two elements of the dimer via the dimer interface (Guo et al. 2005).

Although the availability of cDNAs encoding molecularly defined GPCRs has allowed high-throughput screening for ligands that modulate GPCR function, this is performed almost exclusively in heterologous cell lines transfected to express only the specific GPCR of interest. Given that the human genome contains some 400–450 genes encoding non-chemosensory GPCRs, it is clear that any individual cell of the body may express a considerable number of GPCRs. Interactions between these, either via hetero-dimerisation, via heterologous desensitisation or via the integration of downstream signals can potentially alter the pharmacology, sensitivity and function of receptor agonists and hence produce varied responses. In this article, we will use specific examples to consider the role of homo-dimerisation/oligomerisation in GPCR function and whether either direct hetero-dimerisation or heterologous desensitisation between pairs of co-expressed GPCRs affects the function of the receptor pairs.

1 The Role of α_{1b}-Adrenoceptor Dimerisation/Oligomerisation

Initial studies on potential dimerisation of the α_{1b}-adrenoceptor (Milligan et al. 2004b) employed co-immunoprecipitation of differently epitope-tagged forms of the receptor. When the two forms were co-expressed in heterologous cells, immunoprecipitation with antibody to one of the two epitope tags resulted in the co-immunoprecipitation of the second form (Carrillo et al. 2003; Stanasila et al. 2003; Uberti et al. 2003). This did not occur when different cell populations, each expressing only one form of the receptor, were mixed prior to the immunoprecipitation step. In each of these cases, following SDS-PAGE of the immunoprecipitated samples and immunoblotting to detect the tag not used for immunoprecipitation, a series of bands were detected. The polypeptide with the highest mobility corresponded to the expected size for an α_{1b}-adrenoceptor monomer. Bands with lesser mobility were approximately twice the size of the monomer, whilst a further, less well-defined group of immunoreactive bands entered the gels poorly. Interactions between rhodopsin-like GPCRs are not anticipated to involve SDS-PAGE-resistant covalent links. It thus remains possible that the polypeptides with apparent high molecular mass are simply protein aggregates stemming from removal of the receptors from the membrane lipid environment. In the studies of Carrillo et al. (2003) and Stanasila et al. (2003), further approaches were employed to support the co-immunoprecipitation data. Stanasila et al. (2003) C-terminally tagged the α_{1b}-adrenoceptor with either cyan fluorescent protein (CFP) or green fluorescent protein (GFP) and employed fluorescence resonance energy transfer (FRET) to demonstrate proximity and hence potential interactions when the two forms were co-expressed. They also demonstrated that additional expression of α_{1b}-adrenoceptor not tagged with a fluorescent protein reduced the FRET signal and that such a reduction in FRET signal was not produced by co-expressing the CCR5 chemokine receptor. Carrillo et al. (2003) employed a distinct FRET-based technique. Taking advantage of the N-terminal epitope tags introduced for the co-immunoprecipitation studies, they employed time-resolved FRET between anti-epitope tag antibodies labelled with appropriate energy donor and acceptor species and hence demonstrated interactions between successfully cell surface-delivered forms of the α_{1b}-adrenoceptor in intact cells. Carrillo et al. (2003) also developed a functional complementation strategy. They had previously shown that a single open-reading-frame fusion protein between the α_{1b}-adrenoceptor and the α subunit of the Ca^{2+} mobilising G-protein G_{11} was functional and could be used to measure agonist-stimulated binding of [^{35}S]GTPγS (Carrillo et al. 2002). They extended these studies and showed that the ability of the fusion protein to bind [^{35}S]GTPγS in an agonist-dependent manner could be eliminated by either a point mutation in the second intracellular loop of the receptor or by a point mutation in the G-protein. When the two inactive fusion proteins were co-expressed, agonist-mediated binding of [^{35}S]GTPγS was restored (Carrillo et al. 2003). This also occurred in mouse embryo fibroblasts that lacked endogenous expression of any Ca^{2+} mobilising G-proteins and therefore had to reflect an inter-molecular interaction between the two inactive fusion proteins. Carrillo et al. (2004) subsequently confirmed the capacity of C-terminally CFP and yellow fluorescent protein (YFP)-tagged forms of the α_{1b}-adrenoceptor to generate FRET signals following co-expression.

To attempt to understand the molecular basis of α_{1b}-adrenoceptor dimerisation, Stanasila et al. (2003) generated a series of mutated and modified forms of the receptor. Studies on the yeast α factor receptor had identified a glycophorin A-like motif (GXXXG) in trans-membrane helix I and shown that modification of key amino acids of this motif resulted in a reduction in FRET consistent with abrogation of dimerisation (Overton et al. 2003). Mutation of Gly^{53} within a similar motif in trans-membrane helix I of the α_{1b}-adrenoceptor, however, did not reduce FRET signals (Stanasila et al. 2003). Equivalent mutation of a second glycophorin A-like motif located in trans-membrane helix VI was also without effect. Other studies had suggested roles for both the intracellular C-terminal tail and the glycosylation state of the extracellular N-terminal region in protein–protein interactions of other rhodopsin-like GPCRs (Milligan 2004b). However, neither truncation of the C-terminal tail nor mutation to prevent N-glycosylation reduced FRET signals from suitable α_{1b}-adrenoceptor constructs. Indeed, C-terminal truncation actually resulted in a higher FRET signal (Stanasila et al. 2003). Because resonance energy transfer signals are dependent upon both the distance between and the relative orientation of the energy donor and acceptor (Milligan 2004a; Pfleger and Eidne 2006), interpretation of these observations is difficult. To address the same question, Carrillo et al. (2004) adopted a systematic receptor fragment-interaction approach. This involved taking the extracellular N-terminus of the α_{1b}-adrenoceptor and linking it to fragments of the receptor consisting of trans-membrane domains, I, III, V or VII or I + II, III + IV and V + VI that also incorporated the intracellular loops that link these pairs of trans-membrane domains. Each of these constructs was then used in every possible combination for both co-immunoprecipitation and time-resolved FRET studies to identify interactions. Only trans-membrane domain I and trans-membrane domain IV displayed symmetrical interactions, i.e. they self-associated. Interestingly, both of these regions have been suggested as dimerisation interfaces for other GPCRs (Milligan 2005). However, comparisons with the atomic force microscope images of the in situ organisation of rhodopsin suggested a more complex model (Fig. 1). Viewed from the extra-cellular face of the plasma membrane, the trans-membrane helices of GPCRs are arranged in anti-clockwise orientation. Trans-membrane domain I–trans-membrane domain I interactions between adjacent monomers leaves trans-membrane domain IV available to form a IV–IV interaction with a further monomer, which then leaves trans-membrane domain I free to potentially generate a further interaction (Fig. 1). This pattern could build up into a daisy-chain of repeating dimers to generate an oligomer of undefined size. The fragmentation studies also indicated a series of non-symmetrical interactions involving elements of trans-membrane domains I and/or II with trans-membrane domains V and/or VI and it was hypothesised that such interactions could allow rows of oligomers to form (Carrillo et al. 2004). It is interesting in this regard that near-field scanning optical microscopy has recently imaged large clusters of β-adrenoceptors on both cell lines and murine cardiac myocytes (Ianoul et al. 2005).

Because conventional, two component FRET cannot usefully discriminate between dimers and higher-order structures, J.F. Lopez-Gimenez et al. (personal communication) developed a sequential three-colour FRET imaging approach to gain support for oligomeric structures of the α_{1b}-adrenoceptor in single living cells. Initial, proof of concept, studies employed a single open-reading-frame concatamer of CFP, YFP and DsRed2 to establish appropriate imaging conditions. Following transfection into HEK293 cells, fluorescence corresponding to each of CFP, YFP and DsRed2 could be observed in single cells and, with appropriate excitation and the application of a novel FRET algorithm, each of CFP to YFP, YFP to DsRed2 and CFP to DsRed2 FRET could be measured in single cells. To prove that the CFP to DsRed2 FRET signal truly represented sequential CFP to YFP to DsRed2 energy transfer and not direct CFP to DsRed2 FRET (Fig. 1), a second concatamer was employed. This contained a $Tyr^{67}Cys$ mutation in the YFP element that ablates fluorescence. As such, $Tyr^{67}Cys$ YFP can act as neither a resonance energy transfer donor nor an acceptor. CFP to DsRed2 FRET was

Fig. 1a–d. The α_{1b}-adrenoceptor forms oligomeric chains via interactions involving trans-membrane domains I and IV. **a** Based on interactions between fragments of the α_{1b}-adrenoceptor, where trans-membrane domain I and trans-membrane domain IV (*yellow*) were shown to contribute symmetrical protein–protein interaction interfaces, Carrillo et al. (2004) proposed a daisy-chain structure that may link α_{1b}-adrenoceptor monomers into higher-order oligomers. Cartoons of such oligomers are displayed: *Top*, viewed from the extracellular space, *bottom* viewed as a section through the plasma membrane. Such organisational structure is reminiscent of the arrays of rhodopsin in murine rod outer segments observed by atomic force microscopy (Liang et al. 2003; Fotiadis et al. 2004). **b** The basis of direct CFP-DsRed2 and sequential CFP-YFP-DsRed2 FRET is illustrated. **c** Forms of the wild-type α_{1b}-adrenoceptor (A_{1-6}) and $Leu^{65}Ala$, $Val^{66}Ala$, $Leu^{166}Ala$, $Leu^{167}Ala$ α_{1b}-adrenoceptor (B_{1-6}) were C-terminally tagged with CFP (A_1, B_1), YFP (A_2, B_2) or DsRed2 (A_3, B_3), expressed in HEK293 cells and imaged for CFP, YFP and DsRed2 fluorescence or CFP-YFP (A_4, B_4), YFP-DsRed2 (A_5, B_5) and CFP-DsRed2 (A_6, B_6) FRET. Pseudo-colours are related to FRET intensity. **d** CFP-DsRed2 FRET was quantitated for the co-expressed three protein groups of wild-type α_{1b}-adrenoceptor linked to CFP, YFP and DsRed2 (*black*), wild-type α_{1b}-adrenoceptor linked to CFP, $Tyr^{67}Cys$YFP and DsRed2 (*grey*) and $Leu^{65}Ala$, $Val^{66}Ala$, $Leu^{166}Ala$, $Leu^{167}Ala$ α_{1b}-adrenoceptor linked to CFP, YFP and DsRed2 (*white*). On the right-hand side, the values for the *grey bars* that must represent no sequential FRET (see text for details) have been subtracted. (Data are adapted from Lopez-Gimenez et al., personal communication) eliminated, confirming that the CFP to DsRed2 FRET produced from the initial concatamer represented sequential CFP to YFP to DsRed2 FRET and not direct CFP to DsRed2 FRET. With this information in place, forms of the α_{1b}-adrenoceptor C-terminally tagged with each of CFP, YFP and DsRed2 were co-expressed in HEK293 cells and produced CFP to DsRed2 FRET consistent with the α_{1b}-adrenoceptor existing in an oligomeric complex (Fig. 1), whereas simple co-expression of each of CFP, YFP and DsRed2 produced no such FRET signal. Equally, co-expression of α_{1b}-adrenoceptor-CFP, α_{1b}-adrenoceptor-Tyr^{67}CysYFP and α_{1b}-adrenoceptor-DsRed2 failed to generate CFP to DsRed2 FRET.

To explore the role of oligomerisation of the α_{1b}-adrenoceptor, pairs of adjacent, key hydrophobic residues in trans-membrane domains I and IV were mutated to alanines and three colour FRET studies repeated. The mutations resulted in reduced sequential three-colour FRET

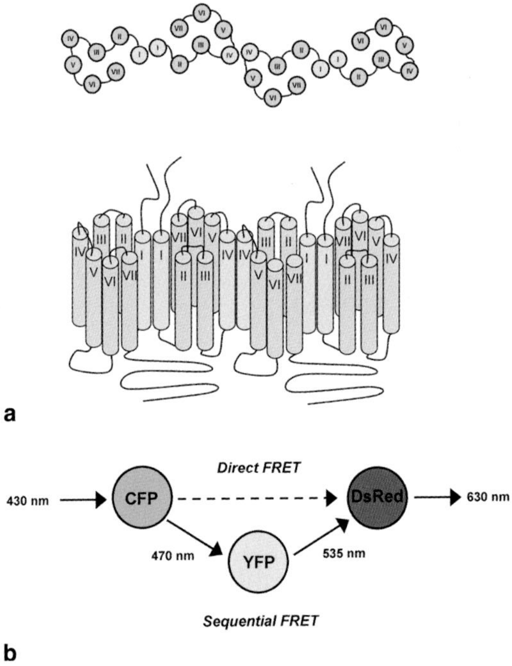

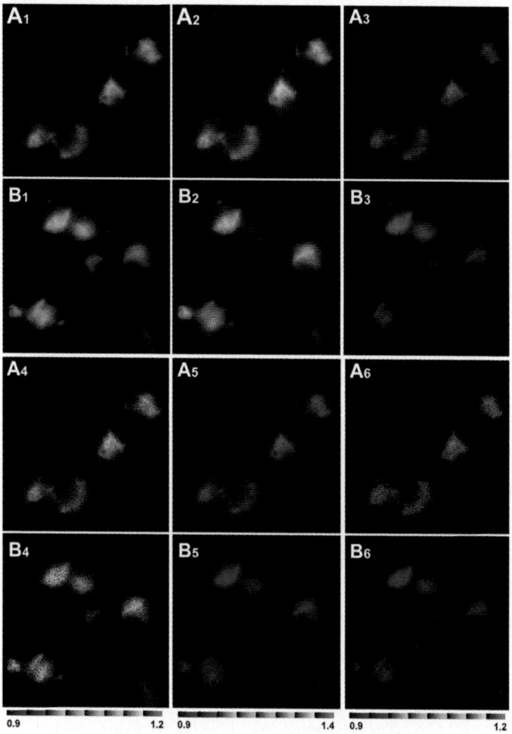
A1
A2
A3
B1
B2
B3
A4
A5
A6
B4
B5
B6
0.9
1.2
0.9
1.4
0.9
1.2

c

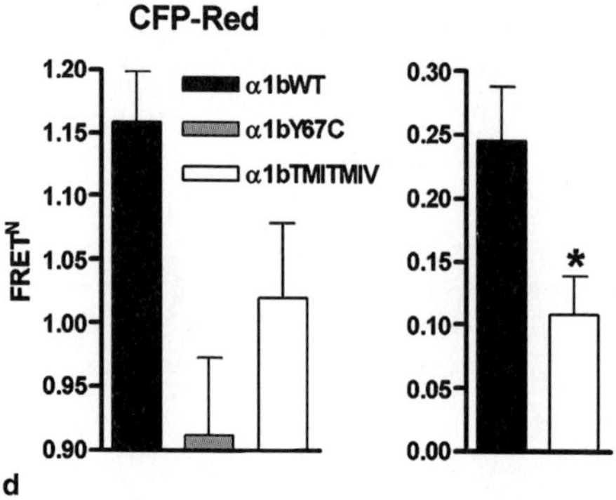
CFP-Red
FRET^N
α1bWT
α1bY67C
α1bTMITMIV
1.20
1.15
1.10
1.05
1.00
0.95
0.90
0.30
0.25
0.20
0.15
0.10
0.05
0.00
*

d

(Fig. 1), consistent with alterations in the oligomeric interactions and organisation of the receptor (J.F. Lopez-Gimenez et al., personal communication). The functional consequences of this were also explored. Immunocytochemistry to detect an N-terminal epitope tag in transfected but nonpermeabilised cells demonstrated α_{1b}-adrenoceptor at the cell surface. However, cell surface expression of the mutant α_{1b}-adrenoceptor could not be detected, although, following cell permeabilisation, it was clear that the mutant was expressed as effectively as the wild-type receptor. Good expression of the mutated form could also be shown by simply monitoring fluorescence of cells expressing C-terminally YFP-tagged forms of the wild-type and mutated α_{1b}-adrenoceptor (J.F. Lopez-Gimenez et al., personal communication).

To explore the basis for the lack of cell surface delivery of the mutated α_{1b}-adrenoceptor, studies were conducted to examine receptor glycosylation. Following SDS-PAGE, the wild-type receptor was present as both apparent 75- and 105-kDa polypeptides, but no 105-kDa form of the mutant was detected. De-glycosylation studies demonstrated the 105-kDa form of the wild-type receptor to represent the mature form of the protein. Following treatment with N-glycosidase, F this band was absent and the wild-type receptor now migrated with apparent M_r close to 70 kDa (J.F. Lopez-Gimenez et al., personal communication). The mutant receptor thus appeared to be unable to become core-glycosylated. It is well established that an inability to be core-glycosylated prevents plasma membrane delivery and trafficking of mutants of a range of GPCRs (Petaja-Repo et al. 2001; Pietila et al. 2005).

As the mutated α_{1b}-adrenoceptor appeared unable to reach the cell surface, then it was expected to be unable to signal in response to agonist ligands. This was confirmed. HEK293 cells expressing either the wild-type α_{1b}-adrenoceptor, C-terminally tagged with YFP, or the mutant C-terminally tagged with CFP were grown on the same coverslip, loaded with the ratiometric Ca^{2+} indicator dye FURA-2 and exposed to the α_1-adrenoceptor agonist phenylephrine. Increases in intracellular Ca^{2+} were only observed in cells expressing the wild-type receptor (J.F. Lopez-Gimenez et al., personal communication). This was not a reflection that the cells expressing the mutated α_{1b}-adrenoceptor were unable to respond to receptor ligands. Subsequent to washout of the α_1-adrenoceptor agonist, ATP was added to activate P_2Y purinoceptors expressed endogenously by HEK 293 cells. Equivalent increases in intracellular Ca^{2+} were then observed in all cells.

All these data indicate not only that the α_{1b}-adrenoceptor is able to form oligomeric, rather than simple dimeric complexes, but also that disruption of effective oligomerisation has profound consequences for receptor maturation and function.

2 Orexin-1 Receptor/Cannabinoid CB1 Receptor Hetero-dimerisation

It is likely GPCRs which are co-expressed and have the capacity to hetero-dimerise may alter the functions of one another. A convenient way to control the timing and extent of expression of a GPCR in the face of constitutive expression of a second GPCR is to take advantage of tetracycline/doxycycline-induced expression from a single defined site of chromosomal integration, as provided by the Flp-In-T-REx HEK293 cell line from Invitrogen. Ellis et al. (2006) initially generated such cells lines harbouring an N-terminally epitope-tagged form of the human orexin-1 receptor that also had in-frame attachment of YFP at the C-terminus. Expression of this protein was completely dependent upon addition of doxycycline. At steady-state in these cells, some 90% of the orexin-1 receptor construct was present at the cell surface. This receptor displayed little evidence of ligand-independent, constitutive activity because both the inducing agent and the peptide agonist orexin A were required to cause phosphorylation of the ERK1 and ERK2 MAP kinases.

As with the vast majority of GPCRs, the sustained presence of agonist resulted in internalisation of the receptor into punctuate, intracellular vesicles. This effect of orexin A was blocked by the orexin-1 receptor antagonist SB-674042 (Langmead et al. 2004) but not by the cannabinoid CB1 receptor antagonist/inverse agonist SR-141716A (Rinaldi-Carmona et al. 1998). A very different pattern of orexin-1 receptor distribution was observed, however, when orexin-1 receptor construct expression was induced in cells that expressed the cannabinoid CB1 receptor constitutively. Now, without addition of orexin A, 50% of the orexin-1 receptor population was in punctuate, intracellular vesicles. To understand the basis of this observation, Ellis et al. (2006) generated further cell lines in which orexin-1 receptor expression could be induced in the face of constitutive expression of a C-terminally CFP-tagged form of the cannabinoid CB1 receptor. In these cells, with or without induction of the orexin-1 receptor, 90% of the cannabinoid CB1-CFP construct was present in recycling, intracellular punctate vesicles. Constitutive recycling of the cannabinoid CB1 receptor has also been described by others, both following expression in HEK293 cells (Leterrier et al. 2004) and for the receptor expressed endogenously in neurones (Leterrier et al. 2006). As in cells expressing the untagged cannabinoid CB1 receptor, induction of orexin-1 receptor resulted in 66% of this receptor being present in intracellular vesicles and pixel-by-pixel analysis indicated very high colour overlap between CFP and YFP fluorescence (Ellis et al. 2006). Not only were there co-expressed GPCRs in vesicles that overlapped, but FRET studies using cannabinoid CB1-CFP as energy donor and orexin-1-YFP as energy acceptor demonstrated their hetero-dimerisation (Ellis et al. 2006). In cells expressing cannabinoid CB1-CFP constitutively, sustained treatment with SR-141716A resulted in a re-distribution of much of the receptor back to the cell surface. Unsurprisingly, treatment with SB-674042 did not alter the distribution of CB1-CFP because this ligand has no inherent affinity to bind the cannabinoid CB1 receptor. Remarkably, however, in cells expressing both orexin-1-YFP and CB1-CFP, treatment with either SR-141716A or SB-674042 returned both GPCRs to the cell surface (Ellis et al. 2006). This is consistent with the FRET experiments that indicated that orexin-1-YFP and CB1-CFP form a hetero-dimer/oligomer.

Hetero-dimerisation/oligomerisation of these two receptors has functional consequences. Sustained treatment of the cells co-expressing the two GPCRs with SR-141716A resulted in a substantial decrease in the potency of orexin-A to stimulate ERK1 and ERK2 phosphorylation. This was not observed following treatment of cells expressing only the orexin-1 receptor with SR-141716A (Ellis et al. 2006). Similarly, treatment of the cells co-expressing these two GPCRs with SB-674042 resulted in decreased potency of the CB1 receptor agonist WIN55,212-2 to cause phosphorylation of ERK1 and ERK2 (Ellis et al. 2006). These studies indicate that, because of GPCR hetero-dimerisation/oligomerisation, effects of ligands in cells and tissues that co-express different GPCRs may involve alterations in pharmacology and function of receptors that the ligand has no direct affinity to bind to. It will be interesting to see how common an effect this is and whether this paradigm may provide insights into unexpected clinical effects of therapeutic molecules.

3 Direct and Indirect Interactions Between the Angiotensin II AT_1 Receptor and the Mas Proto-oncogene

The Mas proto-oncogene is a GPCR and has recently been described as a receptor able to bind the angiotensin II cleavage product Ang 1–7 (Santos et al. 2003) but not authentic angiotensin II. The angiotensin II AT_1 receptor responds to angiotensin II and mediates important biological functions including vasoconstriction, salt/water re-absorption and stimulation of aldosterone release. These two GPCRs are co-expressed in a range of locations, including specific mesenteric blood vessels. Whilst angiotensin II causes concentration-dependent contraction of such blood vessels in mouse, it was noted that the extent of contraction to angiotensin II was substantially greater in vessels taken from Mas knock-out animals (Kostenis et al. 2005). Mas has therefore been described as a physiological antagonist of the AT_1 receptor (Kostenis et al. 2005). At least when co-expressed in heterologous cell systems, Mas and the AT_1 receptor are able to interact directly as monitored by bioluminescence resonance energy transfer and co-immunoprecipitation (Kostenis et al. 2005) and their co-expression results in a decreased ability of angiotensin II to elevate intracellular Ca^{2+} and to stimulate the production of inositol phosphates (Kostenis et al. 2005). However, direct protein–protein interactions may not represent the entire explanation for the observed functional effects. Mas is a proto-oncogene and, as with a number of other GPCRs reported to affect cell division and proliferation (Allen et al. 1991), displays significant levels of constitutive activity (Fig. 2). When expressed along with the Ca^{2+}-mobilising G-protein $G\alpha_{11}$, Mas produced a strong increase in binding of the GTP analogue [^{35}S]GTPγS in $G\alpha_{11}$ immunoprecipitates. By contrast, following mutation of a key hydrophobic amino acid in the second intracellular loop of Mas to produce I^{138}DMas, this receptor variant was unable to

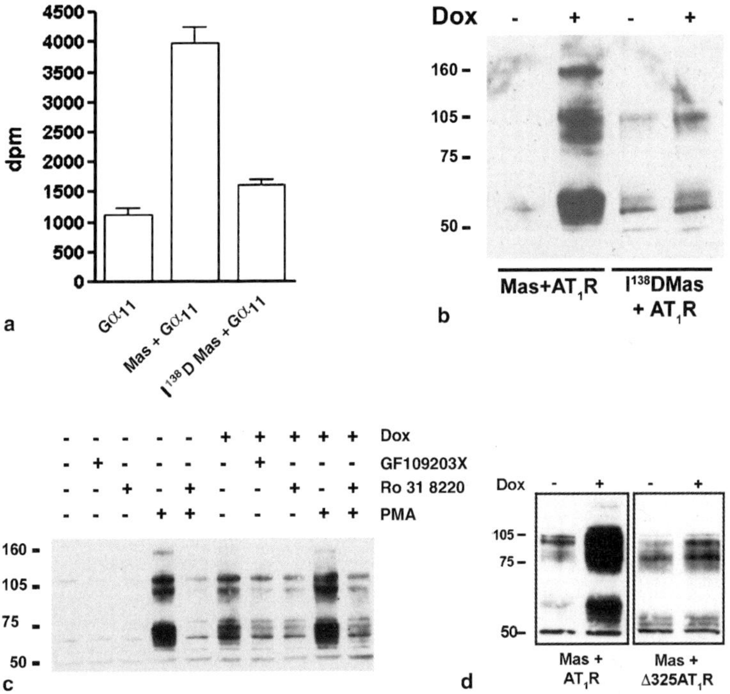

Fig. 2a–d. The constitutive activity of Mas causes up-regulation of a co-expressed angiotensin II AT_1 receptor. **a** The binding of [^{35}S]GTPγS in $G_{11}\alpha$ immunoprecipitates was assessed following expression of $G_{11}\alpha$, Mas + $G_{11}\alpha$, or I^{138}D Mas + $G_{11}\alpha$ in HEK293 cells and membrane preparation. **b** Immunoblots to detect the angiotensin II AT_1 receptor were performed on membranes of Flp-In TREx HEK293 cells constitutively expressing human AT_1 receptor-CFP and with human Mas-YFP or human I^{138}DMas-YFP at the Flp-In locus to allow expression only following treatment with tetracycline/doxycycline (Dox). Strong up-regulation was observed upon induction of Mas-YFP but not I^{138}DMas-YFP. **c** This upregulation is blocked by inhibition of protein kinase C and can be mimicked by activation of protein kinase C without induction of Mas-YFP expression. **d** Mas-induced up-regulation is greatly reduced if the AT_1 receptor lacks protein kinase C consensus sites in the C-terminal tail. A cell line akin to that used in **b** and **c** but constitutively expressing a CFP-tagged form of the AT_1 receptor truncated at amino acid 325 in the C-terminal tail was used in parallel to the cell line employed in **d**. (Data are adapted from Canals et al. 2006)

increase [^{35}S]GTPγS binding to $G\alpha_{11}$. Induced expression of Mas-YFP resulted in increased levels of constitutively expressed AT_1-CFP measured either via [^{3}H]angiotensin II binding, YFP fluorescence or immunoblotting studies (Canals et al. 2006). By contrast, induced expression of I^{138}DMas-YFP did not cause up-regulation of AT_1-CFP. A series of pharmacological inhibitor studies demonstrated that blockade of either $G\alpha_q/G\alpha_{11}$ or protein kinase C (PKC) activity prevented the Mas-induced AT_1-receptor up-regulation, that the effect of Mas could be mimicked by direct activation of PKC and that because the G-proteins lie upstream of PKC, inhibition of $G\alpha_q/G\alpha_{11}$ was unable to prevent the up-regulation produced by direct activation of PKC (Canals et al. 2006).

It has previously been established that the C-terminal tail of the AT_1-receptor contains three consensus sites for PKC-mediated phosphorylation. Canals et al. (2006) thus expressed constitutively a form of AT_1-CFP lacking this region of the C-terminal tail. In these cells, induced expression of Mas-YFP had little ability to up-regulate the AT_1-receptor construct (Fig. 2) (Canals et al. 2006). Because PKC-mediated phosphorylation of the AT_1-receptor has also been associated with functional desensitisation (Smith et al. 1998; Qian et al. 1999), then a desensitised form of the AT_1-receptor, even though present at higher levels, is consistent with the in vivo observation that Mas acts as a physiological brake on AT_1-receptor function and that elimination of Mas would result in more effective responses to angiotensin II and therefore enhanced contraction of mesenteric micro-vessels (Kostenis et al. 2005).

4 Conclusions

The range of studies described here demonstrates the existence of both GPCR homo- and hetero-dimerisation/ologimerisation and indicates that such protein–protein interactions can have major functional consequences. However, with so much research currently exploring aspects of the molecular basis and relevance of GPCR dimerisation/oligomerisation, it is important to remember that the integration of downstream signalling by GPCRs and of heterologous desensitisation of co-expressed GPCRs can also play central roles in defining functional and physiologically relevant endpoints.

References

Allen LF, Lefkowitz RJ, Caron MG, Cotecchia S (1991) G-protein-coupled receptor genes as protooncogenes: constitutively activating mutation of the alpha 1B-adrenergic receptor enhances mitogenesis and tumorigenicity. Proc Natl Acad Sci USA 88:11354–11358

Canals M, Jenkins L, Kellett E, Milligan G (2006) Up-regulation of the angiotensin II AT_1 receptor by the Mas proto-oncogene is due to constitutive activation of G_q/G_{11} by Mas. J Biol Chem 281:16767–16767

Carrillo JJ, Stevens PA, Milligan G (2002) Measurement of agonist-dependent and -independent signal initiation of α_{1b}-adrenoceptor mutants by direct analysis of guanine nucleotide exchange on the G protein Gα11. J Pharmacol Exp Ther 302:1080–1088

Carrillo JJ, Pediani J, Milligan G (2003) Dimers of class A G protein-coupled receptors function via agonist-mediated trans-activation of associated G proteins. J Biol Chem 278:42578–42587

Carrillo JJ, Lopez-Gimenez JF, Milligan G (2004) Multiple interactions between transmembrane helices generate the oligomeric alpha1b-adrenoceptor. Mol Pharmacol 66:1123–1137

Ellis J, Pediani J, Milasta S, Milligan G (2006) Orexin-1 receptor-cannabinoid CB1 receptor hetero-dimerization results in both ligand-dependent and -independent co-ordinated alterations of receptor localization and function. J Biol Chem 281:38812–38824

Fotiadis D, Liang Y, Filipek S, Saperstein DA, Engel A, Palczewski K (2004) The G protein-coupled receptor rhodopsin in the native membrane. FEBS Lett 564:281–288

Guo W, Shi L, Filizola M, Weinstein H, Javitch JA (2005) Crosstalk in G protein-coupled receptors: changes at the transmembrane homodimer interface determine activation. Proc Natl Acad Sci USA 102:17495–17500

Ianoul A, Grant DD, Rouleau Y, Bani-Yaghoub M, Johnston LJ, Pezacki JP (2005) Imaging nanometer domains of beta-adrenergic receptor complexes on the surface of cardiac myocytes. Nat Chem Biol 1:196–202

Kostenis E, Milligan G, Christopoulos A, Sanchez-Ferrer CF, Heringer-Walther S, Sexton P, Gembardt F, Kellett E, Martini L, Vanderheyden P, Schultheiss HP, Walther T (2005) G protein-coupled receptor Mas is a physiological antagonist of the angiotensin II type 1 receptor. Circulation 111:1806–1813

Langmead CJ, Jerman JC, Brough SJ, Scott C, Porter RA, Herdon HJ (2004) Characterisation of the binding of [3H]-SB-674042, a novel nonpeptide antagonist, to the human orexin-1 receptor. Br J Pharmacol 141:340–346

Leterrier C, Bonnard D, Carrel D, Rossier J, Lenkei Z (2004) Constitutive endocytic cycle of the CB1 cannabinoid receptor. J Biol Chem 279:36013–36021

Leterrier C, Laine J, Darmon M, Boudin H, Rossier J, Lenkei Z (2006) Constitutive activation drives compartment-selective endocytosis and axonal targeting of type 1 cannabinoid receptors. J Neurosci 26:3141–3153

Liang Y, Fotiadis D, Filipek S, Saperstein DA, Palczewski K, Engel A (2003) Organization of the G protein-coupled receptors rhodopsin and opsin in native membranes. J Biol Chem 278:21655–21662

Meyer BH, Segura J-M, Martinez KL, Hovius R, George N, Johnsson K, Vogel H (2006) FRET imaging reveals that functional neurokinin-1 receptors are monomeric and reside in membrane microdomains of live cells. Proc Natl Acad Sci USA 103:2138–2143

Milligan G (2004a) Applications of bioluminescence- and fluorescence resonance energy transfer to drug discovery at G protein-coupled receptors. Eur J Pharm Sci 21:397–405

Milligan G (2004b) G protein-coupled receptor dimerization: function and ligand pharmacology. Mol Pharmacol 66:1–7

Milligan G (2005) The molecular basis of dimerisation of family A G protein-coupled receptors. In: Lundstrom K, Chui M (eds) GPCRs in drug discovery. Marcel Dekker, pp 329–340

Milligan G, Bouvier M (2005) Methods to monitor the quaternary structure of G protein-coupled receptors. FEBS J 272:2914–2925

Pietila EM, Tuusa JT, Apaja PM, Aatsinki JT, Hakalahti AE, Rajaniemi HJ, Petaja-Repo UE (2005) Inefficient maturation of the rat luteinizing hormone receptor. A putative way to regulate receptor numbers at the cell surface. J Biol Chem 280:26622–26629

Qian H, Pipolo L, Thomas WG (1999) Identification of protein kinase C phosphorylation sites in the angiotensin II (AT1A) receptor. Biochem J 343:637–644

Rinaldi-Carmona M, Le Duigou A, Oustric D, Barth F, Bouaboula M, Carayon P, Casellas P, Le Fur G (1998) Modulation of CB1 cannabinoid receptor functions after a long-term exposure to agonist or inverse agonist in the Chinese hamster ovary cell expression system. J Pharmacol Exp Ther 87:1038–1047

Santos RA, Simoes e Silva AC, Maric C, Silva DM, Machado RP, de Buhr I, Heringer-Walther S, Pinheiro SV, Lopes MT. Bader M, Mendes EP, Lemos VS, Campagnole-Santos MJ, Schultheiss HP, Speth R, Walther T (2003) Angiotensin-(1–7) is an endogenous ligand for the G protein-coupled receptor Mas. Proc Natl Acad Sci USA 100:8258–8263

Smith RD, Hunyady L, Olivares-Reyes JA, Mihalik B, Jayadev S, Catt KJ (1998) Agonist-induced phosphorylation of the angiotensin AT1a receptor is localized to a serine/threonine-rich region of its cytoplasmic tail. Mol Pharmacol 54:935–941

Stanasila L, Perez JB, Vogel H, Cotecchia S (2003) Oligomerization of the alpha 1a- and alpha 1b-adrenergic receptor subtypes. Potential implications in receptor internalization. J Biol Chem 278:40239–40251

Uberti MA, Hall RA, Minneman KP (2003) Subtype-specific dimerization of alpha 1-adrenoceptors: effects on receptor expression and pharmacological properties. Mol Pharmacol 64:1379–1390

Ernst Schering Foundation Symposium Proceedings, Vol. 2, pp. 163–186
DOI 10.1007/2789_2006_008

Published Online: 16 May 2007

Deorphanization of G-Protein-Coupled Receptors

M. Parmentier(✉), M. Detheux

IRIBHN, ULB Campus Erasme, 808 roude de Lennik, 1070 Bruxelles, Belgium
email: *mparment@ulb.ac.be*

Abstract. G-protein-coupled receptors constitute one of the major families of drug targets. Orphan receptors, for which the ligands and function are still unknown, are an attractive set of future targets for presently unmet medical needs. Screening strategies have been developed over the years in order to identify the natural ligands of these receptors. Natural or chimeric G-proteins that can redirect the natural coupling of receptors toward intracellular calcium release are frequently used. Potential problems include poor expression or trafficking to the cell surface, constitutive activity of the receptors, or the presence of endogenous receptors in the cell types used for functional expression, leading to nonspecific responses. Many orphan receptors characterized over the last 10 years have been associated with previously known bioactive molecules. However, new and unpredicted biological mediators have also been purified from complex biological sources. A few old and recent examples, including nociceptin, chemerin, and the F2L peptide are illustrated. Future challenges for the functional characterization of the remaining orphan receptors include the potential requirement of specific proteins necessary for quality control, trafficking or coupling of specific receptors, the possible formation of obligate heterodimers, and the possibility that some constitutively active receptors may lack ligands or respond only to inverse agonists. Adapted expression and screening strategies will be needed to deal with these issues.

1 G-Protein-Coupled Receptors

G-protein-coupled receptors (GPCRs) are the largest family among the membrane receptors. They play a major role in a variety of physiological and pathophysiological processes, such as carbohydrate metabolism, regulation of the cardiovascular system, nociception, feeding behavior, and immune responses. All GPCRs share a common structural organization with seven transmembrane segments, and a common way of modulating cell function by regulating effector systems through a family of heterotrimeric G-proteins (although G-protein-independent signaling has been reported as well). Not considering the olfactory and gustatory receptors, more than 350 G-protein-coupled receptor types and subtypes have been cloned to date in mammalian species. Among these, approximately 250 have been characterized functionally.

2 Orphan Receptors as Opportunities for Future Drug Targets

Following the cloning of rhodopsin (Nathans and Hogness 1984), β-adrenergic receptors (Dixon et al. 1986; Yarden et al. 1986), and the M1 muscarinic receptor (Kubo et al. 1986), as a result of protein purification and peptide sequencing approaches, the common transmembrane organization and structural relatedness of GPCRs rapidly became clear. As a consequence, polymerase chain reaction using degenerate primers (Libert et al. 1989) over the early 1990s led to the progressive accumulation of a large number of orphan receptors, characterized by a typical GPCR structure, but of unknown function. Later on, the systematic sequencing of cDNA libraries (ESTs) and genomes has further expanded the list of orphan receptors. Due to their accessibility from the extracellular space and their key roles in modulating cell functions, G-protein-coupled receptors constitute the targets for about 40% of the active compounds presently used as therapeutic agents. The pharmaceutical industry is keen on the permanent input of new pharmacological targets in their drug development programs. As G-protein-coupled receptors will certainly remain a major avenue for drug design, characterization of orphan receptors are providing original and attractive targets for therapeutic agents and will likely lead to the development of novel drugs in the future (Ribeiro and Horuk 2005).

3 Expression and Screening Strategies

The identification of the ligands of orphan receptors, starting from purely genetic data, has been referred to as reverse pharmacology. This process is based on the use of specific and sensitive functional assays. As the signaling cascade activated by orphan receptors cannot be predicted for certain, a generic functional assay, independent of the activation of a specific cascade, is generally used. Several of these assays have been proposed and used in the past. Over the last few years, we have used essentially a high-throughput functional assay based on the luminescence emitted by recombinant aequorin following intracellular calcium release (Stables et al. 1997; Le Poul et al. 2002). In this system, a recombinant cell line is developed that coexpresses an orphan receptor, apoaequorin targeted to mitochondria, and $G_{\alpha 16}$ as a generic coupling protein (Fig. 1). Following preincubation of cells with coelenterazine to reconstitute active aequorin, luminescence is recorded in a luminometer following mixing with potential agonists. This assay has been validated with a number of characterized GPCRs and is now used routinely for orphan receptor screening.

Also widely used in the frame of orphan receptor characterization is the classical calcium mobilization assay using fluorescent dyes in a microplate format, following coexpression of the receptor, and $G_{\alpha 15}$, $G_{\alpha 16}$

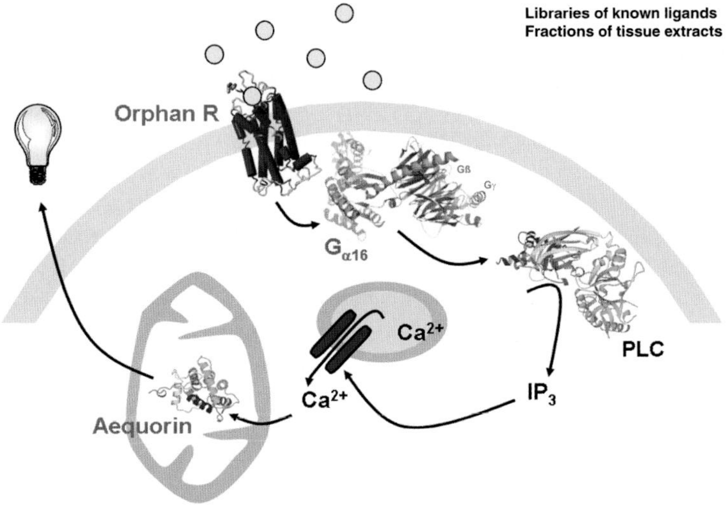

Fig. 1. Aequorin-based assay. The main components of the aequorin-based assay are represented schematically. Three proteins (in *red*) are coexpressed in a CHO-K1 cell line: the orphan receptor, $G_{\alpha16}$ and apoaequorin. $G_{\alpha16}$ allows the coupling of most GPCRs to the phospholipase C (PLC)-IP_3 pathway, irrespective of the natural pathways activated by the receptors. Aequorin is formed by the association of apoaequorin (targeted to mitochondria) and its cofactor coelenterazine. Following receptor activation, the release of Ca^{2+} from intracellular stores results in the activation of aequorin, and the emission of photons recorded by a luminometer. This assay is adapted to 96- and 384-well microplate formats

or hybrid G-proteins (i.e., $G_{\alpha qi5}$) (Offermanns and Simon 1995; Conklin et al. 1993). However, in our hands this technique is less sensitive and less robust than the aequorin-based approach. Other generic techniques include the use of frog melanocytes (Lerner 1994), the internalization of receptor-GFP fusion proteins, or the translocation of a β-arrestin–GFP fusion. Alternatively, cascade-specific assays have been used as well for the deorphanization of specific receptors, including cAMP, GTPγS, and arachidonic acid measurements. These various approaches have been detailed elsewhere (Wise et al. 2004).

All these assays have their specific advantages and limitations, and not all receptors will provide a robust signal in each of them. Using the aequorin-based assay in CHO-K1 cells, we have identified a number of orphan receptors that express poorly in this system, as indicated by particularly low frequencies of clones displaying high transcript levels, the frequent rearrangement of the coding sequence generating the synthesis of nonfunctional receptors, or the low FACS signal obtained on cell lines when monoclonal antibodies are available. Such expression problems are frequently correlated with the demonstration that the receptor displays an apparent constitutive activity. Constitutive activity is usually detected following transient expression of the receptor, and the measurement of cAMP (G_s-coupled receptors), inositol phosphates (G_q-coupled receptors), or GTPγS binding (G_i-coupled), which are recorded as significantly different from the basal levels of untransfected cells. The constitutive activity of the receptor therefore appears as a factor that counter-selects the cell lines expressing it at high levels. We have also determined that some receptors naturally coupled to G_s do not couple efficiently to $G_{\alpha16}$.

Potentially troublesome receptors may be expressed as fusions with a tag, which allow analyzing cell surface expression. As the tag itself may modify the expression or the binding of the receptor ligand, we find it useful to express an untagged receptor in parallel.

We also use an inducible expression vector, based on the tet-on technique. CHO-K1 cell lines coexpressing apoaequorin, $G_{\alpha16}$, and the tet repressor have been established and validated with a number of model receptors. The selected cell line is being used for the expression of the orphan receptors for which constitutive activity and/or other expression problems have been encountered. For each receptor, the doxycycline concentration is adapted by measuring the constitutive activation of intracellular cascades, before screening.

4 Known Molecules and New Biological Mediators

Once established, the cell line expressing an orphan receptor is tested for its functional response to a set of potential ligands. These potential ligands can be well-known biological mediators, for which the precise binding site has not been characterized yet, collections of natural peptides, lipids, or metabolic intermediates with a poorly established role in signaling, or complex biological mixtures. Structural similarities with characterized receptors can of course focus the selection of potential ligands onto specific mediators or chemical classes of potential ligands. Over the years, a large number of orphan receptors have been matched with a well-characterized pharmacology. Other orphan receptors were identified as responding to known ligands, but were characterized either by a novel pharmacology or an original tissue distribution, leading to the multiplication of subtypes in some families, such as the serotonin and chemokine receptors. Finally, a set of orphan receptors were found to respond to previously unknown molecules, as the result of the isolation of these molecules from complex biological sources, on the basis of their biological activity on the recombinant receptor. The first example was the identification of a novel neuropeptide, nociceptin, as the natural agonist of an orphan receptor related to the opioid receptor (see below). Subsequently, orexins, prolactin-releasing peptide, apelin, melanin-concentrating hormone, ghrelin, motilin, urotensin II, prokineticins, kisspeptin/metastin, relaxin-3, and an RFamide peptide have been identified as the natural ligands of previously orphan receptors (Table 1). This is in our view the most attractive side of the orphan receptor field, since naturally processed forms of peptides and proteins, containing necessary tertiary structures and post-translational modifications can be discovered. It is likely that other novel molecules will be similarly discovered in the future, following the analysis of the remaining orphan receptors. With this aim in mind, we are presently expanding our extract preparation and purification schemes, primarily selected to retain peptides and small proteins, in order to focus on other classes of potential ligands, such as bioamines and other small molecules, medium- to large-sized proteins, and lipids. In addition, we also use human clinical samples that have allowed the identification of ligands for some orphan receptors (see below).

Table 1 Natural ligands of human receptors identified through their purification from complex biological sources (tissue extracts of biological fluids)

Receptor	Ligand	Year	Assay	References
ORL1	Nociceptin	1995	cAMP	Meunier et al. 1995
				Reinscheid et al. 1995
HFGAN72	Orexins	1998	Ca^{2+}	Sakurai et al. 1998
APJ	Apelin	1998	Micr.	Tatemoto et al. 1998
GPR10	PrRP	1998	AA	Hinuma et al. 1998
GHSH	Ghrelin	1999	Ca^{2+}	Kojima et al. 1999
GPR14	Urotensin II	1999	Ca^{2+}	Mori et al. 1999
GPR24	MCH	1999	Ca^{2+}	Saito et al. 1999
GPR66	Neuromedin U	2000	Ca^{2+}	Kojima et al. 2000
CCR5	CCL14[9–74]	2000	Ca^{2+}	Detheux et al. 1999
GPR54	Metastin/kisspeptins	2001	Ca^{2+}	Ohtaki et al. 2001
				Kotani et al. 2001
GPR8	NPW	2002	cAMP	Shimonura et al. 2002
				Tanaka et al. 2003
GPR7	NPB	2002	cAMP	Fujii et al. 2002
				Tanaka et al. 2003
GPR73	Prokineticin	2003	cAMP	Lin et al. 2002
ChemR23	Chemerin	2003	Ca^{2+}	Wittamer et al. 2003
GPCR135	Relaxin-3/INSL7	2003	GTPγS	Liu et al. 2003a
GPCR142	Relaxin-3/INSL7	2003	GTPγS	Liu et al. 2003b
GPR91	Succinate	2004	Ca^{2+}	He et al. 2004
GPR154	Neuropeptide S	2004	Ca^{2+}	Xu et al. 2004
FPRL2	F2L	2005	Ca^{2+}	Migeotte et al. 2005
GPR103	QRFP	2006	Luc.	Takayasu et al. 2006

The assay used for the follow-up of their purification is given, together with the year of the reporting in the literature. AA, arachidonic acid assay; Luc., luciferase reporter assay; Micr, microphysiometer

5 ORL1 and Nociceptin

Endogenous opioid peptides are widely distributed in the central and peripheral nervous systems and play important roles in modulating endocrine, cardiovascular, gastrointestinal, and immune functions. Pharmacological studies have defined three classes of opioid receptors termed δ, κ, and μ, which differ in their affinity for various opioid ligands and their distribution in the nervous system (Reisine and Bell 1993). Following the cloning of the δ receptor, reported simultaneously by two groups (Kieffer et al. 1992; Evans et al. 1992), an orphan receptor was cloned by low stringency PCR, and named ORL1 (Mollereau et al. 1994). ORL1 was significantly related to the three classical opioid receptors and to a lesser extent to somatostatin receptors. In situ hybridization demonstrated a large distribution in the central nervous system, distinct from that of opiate receptors. The human recombinant receptor was expressed in CHO-K1 cells, and a large number of natural and synthetic ligands were tested for their potential interaction with ORL1 in binding and functional assays. None of the natural opiate peptides was active, but a functional response (inhibition of forskolin-induced cAMP accumulation) was obtained with high doses of the potent opiate agonist etorphin. The concentrations of etorphin required for the activation of ORL1 (EC_{50} around 1 μM) were two to three orders of magnitude higher than what is necessary to achieve a similar effect on opiate receptors (Mollereau et al. 1994). These results demonstrated, however, that ORL1 was coupled, like opiate receptors, to the inhibition of adenylyl cyclase, and that the cell line expressing the orphan receptor could be used as a functional assay to detect the activity of agonists. The cell line expressing human ORL1 was therefore used as a bioassay to detect biological activities in extracts from rat brain. Following a gel filtration step, a fraction was found to be active, and the biological activity was purified to homogeneity by FPLC and HPLC. The active compound was characterized by mass spectrometry as a novel heptadecapeptide, FGGFTGARKSARKLANQ, sharing similarity with the endogenous opioid peptide dynorphin A (Meunier et al. 1995). The synthetic peptide exhibits nanomolar potency in inhibiting forskolin-induced accumulation of cAMP. When administered intra-cerebro-ventricularly in mice, the peptide was shown to induce hyperalgia in a hot plate assay, and was therefore termed nociceptin (Meunier et al. 1995). The same peptide was isolated independently by Reinscheid et al. (1995) and named orphanin FQ. The prepronociceptin (pP-NOC) gene displays organizational and structural features that are very similar to those of the genes encoding the precursors to endogenous opioid peptides, enkephalins (pPENK), dynorphins/neo-endorphins (pP-

DYN), and β-endorphin (pPOMC), demonstrating its evolution from a common ancestor (Mollereau et al. 1996). Pronociceptin contains cleavage sites suggesting the generation of other potentially bioactive peptides. A C-terminal peptide of 28 amino acids, whose sequence is strictly conserved across murine and human species, was later described as nocistatin, displaying analgesic properties in vivo (Okuda-Ashitaka et al. 1998).

Nociceptin has since been described to display a range of activities, as a consequence of the broad distribution of ORL1 in the central nervous system. Nociceptin can exhibit both antiopiate as well as analgesic properties, depending on the experimental setting and its site of action (Heinricher 2005). The nociceptin-ORL1 system is now considered as a target for the development of drugs in the fields of pain, anxiety, drug dependence, and obesity (Zaveri et al. 2005; Reinscheid 2006).

6 Characterization of Chemerin as the Natural Ligand of ChemR23

A large number of G-protein-coupled receptors contribute to the mounting of immune responses by regulating the trafficking of leukocyte populations. Chemokines constitute one of the major classes of signaling proteins in this frame (Rossi and Zlotnik 2000; Sallusto et al. 2000), with over 40 chemokines and 19 chemokine receptors described so far (Murphy et al. 2000). Other chemoattractant molecules include the formyl peptides, complement fragments (C3a, C5a), and leukotrienes, among others. A number of orphan human receptors are structurally related to chemoattractant receptors.

ChemR23 is a receptor that was initially described to be expressed in immature dendritic cells (DCs) and macrophages (Samson et al. 1998). Given this distribution pattern and the rapid down-modulation following maturation of DCs, we speculated that the ligand was generated in inflammatory conditions. We therefore used the receptor in a bioassay, and tested fractions derived from human inflammatory samples. A biological activity, specific for ChemR23, was identified in a human ascitic fluid secondary to an ovarian carcinoma (Wittamer et al. 2003). The purification of this activity led to the characterization of the bioactive molecule as the product of *TIG-2* (tazarotene-induced gene-2), a gene previously shown to be induced in keratinocytes by analogs of vitamin A, and overexpressed in patients with psoriasis (Nagpal et al. 1997). This natural ligand of ChemR23 was named chemerin. Chemerin is structurally related to the cathelicidin precursors (antibacterial peptides), cystatins (cysteine protease inhibitors), and kininogens (Fig. 2). Like other chemoattractant receptors, chemerin was shown to act through the G_i class of G-proteins.

Chemerin is synthesized as a secreted precursor, prochemerin, which is poorly active, but converted into a full agonist of ChemR23 by the proteolytic removal of the last six or seven amino acids. We have determined that a synthetic nonapeptide corresponding to the C-terminal end of mature chemerin is able to activate the receptor with limited loss of potency as compared to the full-size protein (Wittamer et al. 2004). Neutrophil cathepsin G and elastase were identified as two proteases able to activate prochemerin, generating two chemerin forms differing by a single amino acid at their C-terminus (Wittamer et al. 2005). Enzymes of the coagulation cascades have been described as processing

Fig. 2a,b. Structure of chemerin. **a** The amino acid sequence of human preprochemerin (*Chem*) is aligned with other proteins containing a cystatin fold. This includes the precursors of the human cathelicidin FALL39 (*FA39*), the mouse Cramp (*CRAM*) and porcine protegrin (*PTG*), the first domain of bovine kininogen (*KNNG*), and the chicken egg-white cystatin (*CYST*). The signal peptides are represented in *lowercase italics*. The cysteines involved in disulfide bonding (of which four are conserved across the family) are in *green*. The *red arrowheads* indicate (when known) the position of the introns in the structure of the respective genes. In all cases, the introns interrupt the coding sequences between codons. C-terminal peptides that are cleaved by proteolysis are represented in *blue*. This results in the generation of active chemerin (N-terminal domain), while for cathelicidin precursors, the released C-terminal peptides display bactericidal properties. The nonapeptide represented in *red* and underlined is chemerin-9, the synthetic peptide that binds and activates the ChemR23 receptor with low nanomolar potency. **b** Tridimensional structure of the cystatin-like domain of porcine protegrin, a cathelicidin precursor (PDB 1PFP). Chemerin is expected to adopt a similar conformation prochemerin as well (Zabel et al. 2005). The receptor was also shown to recruit both myeloid and plasmacytoid dendritic cells (Vermi et al. 2005). Resolvin E1, an omega-3 lipid mediator, was also proposed recently as a ligand for ChemR23 (Arita et al. 2005), although it is not yet clear whether the anti-inflammatory activities of resolvin E1 are indeed mediated through ChemR23.

```
hChem            mrrlliplalwlgavgvGVAELTEAQRRGLQVALEEFHKHPPVQWAFQETS
hFA39 mktqrdghslgrwslvllllglvmplaiiaQVLSYKEAVLRAIDGINQRSSDANLYRLLD
mCRAM    mqfqrdvpslwlwrslslllllglgfsQTPSYRDAVLRAVDDFNQQSLDTNLYRLLD
pPTG   metqraslclgrwslwlllalvvpsasaQALSYREAVLRAVDRLNEQSSEANLYRLLE
bKNNG      mklitilflcsrllpsltQESSQEIDCNDQDVFKAVDAALTKYNSENKSGNQFVLY
cCYST magarg...vgavlgSEDRSRLLGAPVPVDENDEGLQRALQFAMAEYNRASNDKYSSRVV

Chem VESAVDTPFPAGIFVRLEFKLQQTSCRKRDWKKPE-CKVRP---NGRKRKCLACIKLG-SED
FA39 LDPRPTMDGDPDTPKPVSFTVKETVCPRTTQQSPEDCDFKK---DGLVKRCMGTVTLN---Q
CRAM LDPEPQGDEDPDTPKSVRFRVKETVCGKAERQLPEQCAFKE---QGVVKQCMGAVTLN---P
PTG  LDQPPKADEDPGTPKPVSFTVKETVCPRPTRQPPELCDFKE---NGRVKQCVGTVTLD---Q
KNNG RITEVARMDNPDTFYSLKYQIKEGDCPFQSNKTWQDCDYKD-SAQAATGQCTATVAKR-GNM
CYST RVISAKRQLVSGIKYILQVEIGRTTCPKSS-GDLQSCEFHDEPEMAKYTTCTFVVYSIPWLN

Chem KVLGRLVHCPIETQVLREAEEHQETQCLRVQRAGEDPSFYFPGQFAFSKALPRS
FA39 ARGSFDISCDKDNKRFALLGDFFRKSKEKIGKEFKRIVQRIKDFLRNLVPRTES
CRAM AADSFDISCNEPGAQPFRFKKISRLAGLLRKGGEKIGEKLKKIGQKIKNFFQKLVPQPEQ
PTG  IKDPLDITCNEVQGVRGGGLCYCRRRFCVCVGRG
KNNG KFSVAIQTCLITPAEGPVVTAQY....
CYST QIKLLESKCQ
```

a

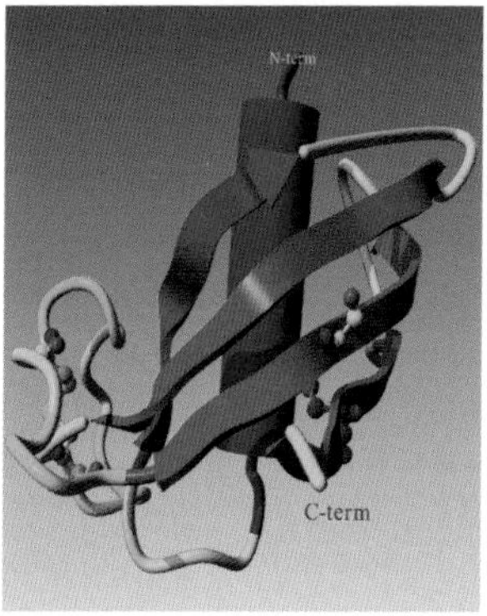

b

Clear orthologs can be found in rodents for both the ChemR23 receptor and prochemerin. The binding and functional parameters of the mouse system were very similar to those observed in human (K_D and EC_{50} around 1 nM for chemerin), with full cross-reactivity between the human and mouse components.

Therefore, chemerin appears as a potent chemoattractant of a novel class. Expression of ChemR23 is essentially restricted to macrophages and dendritic cells and its agonist chemerin can be found at a high level in human inflammatory fluids. We therefore postulate that chemerin is involved in the recruitment of APCs and regulates the inflammatory process and the development of adaptive immune responses. The characterization of a knock-out model for ChemR23 is ongoing and will make it possible to determine the phenotype associated with ChemR23 deficiency and its involvement in disease.

7 Characterization of F2L and Humanin as High-Affinity Endogenous Agonists of FPRL2

FPRL2 belongs to the family of receptors similar to FPR, the receptor for formylated peptides of bacterial origin. Formyl peptide receptors play an essential role in host defense mechanisms against bacterial infection and in the regulation of inflammatory reactions. In human, this family includes three receptors: FPR, FPRL1, and FPRL2. FPR is expressed in neutrophils, monocytes, and DCs, FPRL1 in neutrophils and monocytes, and FPRL2 in monocytes and DCs. FPRL1 is a promiscuous receptor, responding to a large variety of ligands of endogenous and exogenous origins and high structural diversity, including lipoxin A4, serum amyloid 4, and bacterial peptides (Le et al. 2001, 2002). Many of these ligands are low-affinity ligands, and their functional relevance is therefore questionable. A few of these FPRL1 agonists were found to activate FPRL2 at high (μM) concentrations as well, but no high-affinity ligands have been described so far for this receptor. Using a CHO-K1 cell line expressing human FPRL2, $G_{\alpha 16}$ and apoaequorin, we tested fractions of tissue extracts for biological activities, focusing on lymphoid organs. Two specific activities were found in fractions of human and porcine spleen, and the bioactive porcine compounds were purified to homogeneity. These peptidic ligands were characterized as N-terminal peptides of the intracellular heme-binding protein (HBP). The most active peptide was an acetylated 21 amino acid peptide (Ac-MLGMIKNSLFGSVETWPWQVL), perfectly conserved between human and pig, and was named F2L (Fig. 3). It displayed a 5- to 10-nM EC_{50} for human FPRL2 according to the assay. Acetylation of the peptide is not essential for its activity. The second activity corresponded to a longer peptide, incompletely characterized, but presenting a much lower potency. F2L was demonstrated to trigger intracellular calcium release, inhibition of cAMP accumulation, and phosphorylation of ERK1/2 MAP kinases through the G_i class of G-proteins in FPRL2-expressing cells. It activates and chemoattracts monocytes and immature dendritic cells. F2L is inactive on FPR, and poorly active on FPRL1 (Migeotte et al. 2005). HBP, described as binding various molecules containing a tetrapyrrole structure, is poorly characterized functionally (Taketani et al. 1998; Jacob Blackmon et al. 2002). F2L therefore appeared as a new natural chemoattractant peptide for DCs and monocytes in human, and the first potent and specific agonist of FPRL2. In parallel, another human endogenous peptide, humanin, was described as a high-affinity ligand of FPRL2, but this ligand is shared with FPRL1 (Harada et al. 2004).

As HBP is an intracellular protein, it is unclear at this stage what mechanism is involved in the release of the F2L peptide. Our hypothesis is that cell death by apoptosis or necrosis would be responsible for the proteolytic generation of F2L from HBP and its release in the extracellular space. This hypothesis is presently being tested.

8 Future Challenges

Besides olfactory receptors, approximately 100 GPCRs presently remain orphan. A number of associations between ligands and orphan receptors proposed over the recent years are still a matter of debate, and it is likely that some receptors presently considered deorphanized will find more convincing agonists in the future. It is also likely that the easy ligands have been identified, and that the remaining set of orphan receptors concentrates a number of obstacles that will make their characteri-

```
Human                MLGMIKNSLFGSVETWPWQVLSKGDKEEVAYEERACEGGKFATVEVTDKPVDEALREAMPKVA
Pig                  MLGMIKNSLFGSVETWPWQVLSKGDKQDISYEERACEGGKFATVEVTDKPVDEALREAMPKVM
Mouse                MLGMIRNSLFGSVETWPWQVLSTGGKEDVSYEERACEGGKFATVEVTDKPVDEALREAMPKIM
Rat                  MLGMIRNSLFGSVETWPWQVLSTGGKEDVSYEERACEGGKFATVEVTDKPVDEALREAMPKIM
Chicken              MLGMIKNSLLSTVETWPYRVLSKGEKEQLSYEERECEGGQFAVVEVTGKPFDEASKEAALKLL
Xenopus              MFGMIKNSLLGGVENNEGKLVSKGEKDGVAFEEREYEGGKFISTEVSGKPFDEASKEGVLRLL
Tetraodon            MFGMIKNSLFGNTEKTEYKLLSSETKDGVSFEVRRYDAAKYATVSSEGRTFDQISGELVRKLL
Soul (Hum) MAEPLQPDPGAAEDAAAQAVETPGWKAPEDAGPQPGSYEIRHYGPAKWVSTSVESMDWDSAIQTGFTKLN

Human      KYAGGTNDKGIGMGMTVPISFAVFPNEDGSLQKKLKVWFRIPNQFQSDPPAPSDKSVKIEEREGITVYSM
Pig        KYVGGSNDKGIGMGMTVPISFAVFPSDGGSLQKKLKVWFRIPNEFQSNPPVPSDDSIKIEERESITVYSL
Mouse      KYVGGTNDKGVGMGMTVPVSFAVFPNEDGSLQKKLKVWFRIPNQFQGSPPAPSDESVKIEEREGITVYST
Rat        KYVGGTNDKGVGMGMTVPISFAVFPNEDGSLQKKLKVWFRIPNQFQGSPPTPSDQSVKIEEREGITVYST
Chicken    KYVGGSNDKGTGMGMTAPVSITAFPAADGSLQQKVKVYLRIPNQFQASPPCPSDESIKIEERQGMTIYST
Xenopus    KYVGGSNNKSAGMGMTSPVIINSYPSENDTLQPNVKVLLRIPSQYQADPPVPTDNTIQIEDRESVTLYST
Tetraodon  MYIGGSNEQGEAMGTATPI--------------------------YQQSPPTPSDTAVKIEERPGMTVYAL
Soul (Hum) SYIQGKNEKEMKIKMTAPVTSYVEPGSGPFSESTITISLYIPSEQQFDPPRPLESDVFIEDRAEMTVFVR

Human      QFGGYAKEADYVAQATRLRAALEGT-ATYRGDIYFCTGYDPPMKPYGRRNEIWLLKT
Pig        QFGGYAKEADYVARAAQLRTALEGI-ATCRSDVYFCTGYDPPMKPYGRRNEVWLVKA
Mouse      QFGGYAKEADYVAHATQLRTTLEGTPATYQGDVYYCAGYDPPMKPYGRRNEVWLVKA
Rat        QFGGYAKEADYVAHATQLRTTLEGTPATYQGDVYYCAGYDPPMKPYGRRNEVWLVKV
Chicken    QFGGYAKEVDYVNYAAKLKTAL-GSEAAYRKDFYFCNGYDPPMKPYGRRNEVWFVKE
Xenopus    [illegible]
Tetraodon  QFGGFAGESEYRAEALRLTRTL-GETAPYQRKQYFCCSYDPPLKPYGRCNEVWFLQDEP
Soul (Hum) SFDGFSSAQKNQEQLLTLASILREDGKVFDEKVYYTAGYNSPVKLLNRNNEVWLIQKNEPTKENE
```

Fig. 3. The F2L peptide and its HBP precursor. Alignment of the heme-binding protein (HBP) from various species, together with the most closely related human protein, SOUL. Identities with the human HBP sequence are represented in *red*. The underlined peptide in *blue* is F2L, the N-terminally acetylated peptide isolated from porcine spleen as a natural agonist of human FPRL2. The sequence of F2L is identical in human, and well conserved (as the remaining part of HBP) in more distant species

zation difficult. Indeed, all orphan receptors have been tested by numerous groups for their response to large collections of bioactive molecules; they have also been tested for their response to tissue extracts, particularly peptidic extracts. There is still a set of ligands that have not found their receptor, including CART, nocistatin, motilin-associated peptide, neuropeptide GE, neuropeptide EI, NocII, BRAK, and EMAPII among others. However, it is likely that the ligands of most of the remaining orphan receptors are unknown biological mediators that are either unstable, expressed at low levels or in restricted regions, tightly regulated in specific situations, or only present at precise developmental stages. Our experience and the experience of others have shown that many new ligands could not be predicted from standard genomic analysis, as the active compounds required post-translational modifications such as proteolytic processing or grafting of lipid moieties. This suggests that many of the ligands expected for orphan receptors will require their purification from complex biological samples.

Another potential issue that may hinder the characterization of the remaining orphan receptors is the potential requirement for protein partners that may affect their folding, trafficking, pharmacology, or the efficiency of their coupling to signal transduction cascades. The first example of this type was RAMPs, single-pass membrane proteins that determine which agonists are able to activate the complex, but are also required for its traffic from the endoplasmic reticulum to the plasma membrane (McLatchie et al. 1998). The association of RAMP1 with the calcitonin-receptor-like receptor (CRLR) results in a CGRP receptor, while the association of RAMP2 with the same GPCR results in an adrenomedullin receptor. Although the RAMP family is limited and influences the function of a restricted number of receptors (Parameswaran and Spielman 2006), various proteins belonging to diverse structural and functional families have been shown over the years to influence GPCR properties. The MC2 (ACTH) receptor was reported to require another single-pass transmembrane protein, MRAP, for its trafficking to the cell surface (Metherell et al. 2005). Olfactory receptors, which for years have proven to be extremely tedious to express functionally in classical heterologous systems, have been shown to require chaperones such as RTP1, RTP2, and REEP1 in order to traffic properly to the plasma membrane (Saito et al. 2004).

G-protein-coupled receptors have been shown over recent years to form homo- and heterodimers (Bulenger et al. 2005). Some receptors require the heterodimerization of two different polypeptides in order to form functional receptors. The first well-established example is the $GABA_B$ receptor, for which the $GABA_{B1}$ polypeptide, which is able to bind GABA, contains a ER-retention signal that prevents its trafficking to the plasma membrane. The association with the $GABA_{B2}$ polypeptide, which does not bind GABA, masks the retention signal and allows the trafficking of the heterodimer and its functional response (Pin et al. 2003). A similar situation is found in taste receptors. A common $T1R_1$ polypeptide forms a sweet receptor when associated with $T1R_2$ and a L-amino acid sensor when associated with $T1R_3$ (Nelson et al. 2001, 2002). Obligate heterodimers have also been described for insect olfactory receptors (Benton et al. 2006), although this property does not seem to be shared by mammalian olfactory receptors. Besides obligate heterodimers, there is growing evidence that heterodimerization of GPCRs that are fully functional when expressed alone can modify the pharmacology of the partners involved. This was first demonstrated for opiate receptors, for which heterodimers can form original binding sites (Jordan and Devi 1999). We have demonstrated that heterodimerization of chemokine receptors could also modify the pharmacology of both receptors through an allosteric interaction between the binding sites of each protomer (Springael et al. 2005, 2006). It is therefore possible that some orphan receptors might be functional only when coexpressed as heterodimers, or alternatively that the only function of an orphan GPCR might be to modify the pharmacology of a presently characterized receptor. Characterization of these potential situations will require detailed analyses of expression patterns, in order to determine which receptors are coexpressed in each cell type, and raise hypotheses regarding possible heterodimers.

A last set of conditions that may render the characterization of orphan receptors troublesome is that some of them may exhibit signaling properties different from the classical pathways activated by GPCRs. It is now well established that GPCRs may activate intracellular cascades independently from G proteins (Luttrell 2006). There is so far no well-established example of a GPCR acting solely through G-protein-independent pathways, but this possibility must be considered. Also, many receptors display constitutive activities when overexpressed, and some, including a number of orphans, keep such constitutivity when expressed at physiological levels. This suggests that some of these receptors might regulate the cells as a result of their expression, without the need for agonists. Alternatively, some constitutively active receptors may be regulated by natural inverse agonists, rather than by agonists. This might be considered as an example for GPR3, displaying strong constitutivity toward the adenylyl cyclase pathway (Eggerickx et al. 1995), and involved in the blockade of the cell cycle during oogenesis (Mehlmann et al. 2004; Ledent et al. 2005). Finally, some receptors might bind ligands without promoting signaling in the cell. Such behavior has been convincingly proposed for several receptors belonging to the chemokine receptor family, such as DARC and D6. These receptors are considered as decoy receptors, able to bind a diverse set of chemokines, internalize them, and drive them to degradation, thereby contributing to the dampening of inflammatory responses (Middelton et al. 2002; Locati et al. 2005). These receptors have also been proposed as promoting transcytosis of chemokines across endothelial cells. Each of these situations will require the design of specific assays in order to test the various hypotheses specifically.

Acknowledgements. The research conducted in the authors' laboratory was supported by the Actions de Recherche Concertées of the Communauté Française de Belgique, the French Agence Nationale de Recherche sur le SIDA, the Belgian programme on Interuniversity Poles of attraction initiated by the Belgian State, Prime Minister's Office, Science Policy Programming, the European Union (grants LSHB-CT-2003–503337/GPCRs and LSHB-CT-2005–518167/INNOCHEM), the Fonds de la Recherche Scientifique Médicale of Belgium, and the Fondation Médicale Reine Elisabeth. The scientific responsibility is assumed by the authors.

References

Arita M, Bianchini F, Aliberti J, Sher A, Chiang N, Hong S, Yang R, Petasis NA, Serhan CN (2005) Stereochemical assignment, antiinflammatory properties, and receptor for the omega-3 lipid mediator resolvin E1. J Exp Med 201:713–722

Benton R, Sachse S, Michnick SW, Vosshall LB (2006) Atypical membrane topology and heteromeric function of Drosophila odorant receptors in vivo. PLoS Biol 4:e20

Bulenger S, Marullo S, Bouvier M (2005) Emerging role of homo- and heterodimerization in G-protein-coupled receptor biosynthesis and maturation. Trends Pharmacol Sci 26:131–137

Conklin BR, Farfel Z, Lustig KD, Julius D, Bourne HR (1993) Substitution of three amino acids switches receptor specificity of Gq alpha to that of Gi alpha. Nature 363:274–276

Detheux M, Standker L, Vakili J, Munch J, Forssmann U, Adermann K, Pohlmann S, Vassart G, Kirchhoff F, Parmentier M, Forssmann WG (2000) Natural proteolytic processing of hemofiltrate CC chemokine 1 generates a potent CC chemokine receptor (CCR)1 and CCR5 agonist with anti-HIV properties. J Exp Med 192:1501–1508

Dixon RA, Kobilka BK, Strader DJ, Benovic JL, Dohlman HG, Frielle T, Bolanowski MA, Bennett CD, Rands E, Diehl RE, Mumford RA, Slater EE, Sigal IS, Caron MG, Lefkowitz RJ, Strader CD (1986) Cloning of the gene and cDNA for mammalian beta-adrenergic receptor and homology with rhodopsin. Nature 321:75–79

Eggerickx D, Denef JF, Labbe O, Hayashi Y, Refetoff S, Vassart G, Parmentier M, Libert F (1995) Molecular cloning of an orphan G-protein-coupled receptor that constitutively activates adenylate cyclase. Biochem J 309:837–843

Evans CJ, Keith DE Jr, Morrison H, Magendzo K, Edwards RH (1992) Cloning of a delta opioid receptor by functional expression. Science 258:1952–1955

Fujii R, Yoshida H, Fukusumi S, Habata Y, Hosoya M, Kawamata Y, Yano T, Hinuma S, Kitada C, Asami T, Mori M, Fujisawa Y, Fujino M (2002) Identification of a neuropeptide modified with bromine as an endogenous ligand for GPR7. J Biol Chem 277:34010–34016

Harada M, Habata Y, Hosoya M, Nishi K, Fujii R, Kobayashi M, Hinuma S (2004) N-Formylated humanin activates both formyl peptide receptor-like 1 and 2. Biochem Biophys Res Commun 324:255–261

He W, Miao FJ, Lin DC, Schwandner RT, Wang Z, Gao J, Chen JL, Tian H, Ling L (2004) Citric acid cycle intermediates as ligands for orphan G-protein-coupled receptors. Nature 429:188–193

Heinricher MM (2005) Nociceptin/orphanin FQ: pain, stress and neural circuits. Life Sci 77:3127–3132

Hinuma S, Habata Y, Fujii R, Kawamata Y, Hosoya M, Fukusumi S, Kitada C, Masuo Y, Asano T, Matsumoto H, Sekiguchi M, Kurokawa T, Nishimura O, Onda H, Fujino M (1998) A prolactin-releasing peptide in the brain. Nature 393:272–276

Jacob Blackmon B, Dailey TA, Lianchun X, Dailey HA (2002) Characterization of a human and mouse tetrapyrrole-binding protein. Arch Biochem Biophys 407:196–201

Jordan BA, Devi LA (1999) G-protein-coupled receptor heterodimerization modulates receptor function. Nature 399:697–700

Kieffer BL, Befort K, Gaveriaux-Ruff C, Hirth CG (1992) The delta-opioid receptor: isolation of a cDNA by expression cloning and pharmacological characterization. Proc Natl Acad Sci USA 89:12048–12052

Kojima M, Hosoda H, Date Y, Nakazato M, Matsuo H, Kangawa K (1999) Ghrelin is a growth-hormone-releasing acylated peptide from stomach. Nature 402:656–660

Kojima M, Haruno R, Nakazato M, Date Y, Murakami N, Hanada R, Matsuo H, Kangawa K (2000) Purification and identification of neuromedin U as an endogenous ligand for an orphan receptor GPR66 (FM3). Biochem Biophys Res Commun 276:435–438

Kotani M, Detheux M, Vandenbogaerde A, Communi D, Vanderwinden JM, Le Poul E, Brezillon S, Tyldesley R, Suarez-Huerta N, Vandeput F, Blanpain C, Schiffmann SN, Vassart G, Parmentier M (2001) The metastasis-suppressor gene KiSS-1 encodes kisspeptins, the natural ligands of the orphan G protein-coupled receptor GPR54. J Biol Chem 276:34631–34636

Kubo T, Fukuda K, Mikami A, Maeda A, Takahashi H, Mishina M, Haga T, Haga K, Ichiyama A, Kangawa K, Kojima M, Matsuo H, Hirose T, Numa S (1986) Cloning, sequencing and expression of complementary DNA encoding the muscarinic acetylcholine receptor. Nature 323:411–416

Le Y, Oppenheim JJ, Wang JM (2001) Pleiotropic roles of formyl peptide receptors. Cytokine Growth Factor Rev 12:91–105

Le Y, Murphy PM, Wang JM (2002) Formyl-peptide receptors revisited. Trends Immunol 23:541–548

Le Poul E, Hisada S, Mizuguchi Y, Dupriez VJ, Burgeon E, Detheux M (2002) Adaptation of aequorin functional assay to high throughput screening. J Biomol Screen 7:57–65

Ledent C, Demeestere I, Blum D, Petermans J, Hamalainen T, Smits G, Vassart G (2005) Premature ovarian aging in mice deficient for Gpr3. Proc Natl Acad Sci USA 102:8922–8926

Lerner MR (1994) Tools for investigating functional interactions between ligands and G-protein-coupled receptors. Trends Neurosci 17:142–146

Libert F, Parmentier M, Lefort A, Dinsart C, Van Sande J, Maenhaut C, Simons MJ, Dumont JE, Vassart G (1989) Selective amplification and cloning of four new members of the G protein-coupled receptor family. Science 244:569–572

Lin DC, Bullock CM, Ehlert FJ, Chen JL, Tian H, Zhou QY (2002) Identification and molecular characterization of two closely related G protein-coupled receptors activated by prokineticins/endocrine gland vascular endothelial growth factor. J Biol Chem 277:19276–19280

Liu C, Chen J, Sutton S, Roland B, Kuei C, Farmer N, Sillard R, Lovenberg TW (2003a) Identification of relaxin-3/INSL7 as a ligand for GPCR142. J. Biol Chem 278:50765–50770

Liu C, Eriste E, Sutton S, Chen J, Roland B, Kuei C, Farmer N, Jornvall H, Sillard R, Lovenberg TW (2003b) Identification of relaxin-3/INSL7 as an endogenous ligand for the orphan G-protein-coupled receptor GPCR1J. Biol Chem 278:50754–50764

Locati M, Torre YM, Galliera E, Bonecchi R, Bodduluri H, Vago G, Vecchi A, Mantovani A (2005) Silent chemoattractant receptors: D6 as a decoy and scavenger receptor for inflammatory CC chemokines. Cytokine Growth Factor Rev 16:679–686

Luttrell LM (2006) Transmembrane signaling by G protein-coupled receptors. Methods Mol Biol 332:3–49

McLatchie LM, Fraser NJ, Main MJ, Wise A, Brown J, Thompson N, Solari R, Lee MG, Foord SM (1998) RAMPs regulate the transport and ligand specificity of the calcitonin-receptor-like receptor. Nature 393:333–339

Mehlmann LM, Saeki Y, Tanaka S, Brennan TJ, Evsikov AV, Pendola FL, Knowles BB, Eppig JJ, Jaffe LA (2004) The Gs-linked receptor GPR3 maintains meiotic arrest in mammalian oocytes. Science 306:1947–1950

Metherell LA, Chapple JP, Cooray S, David A, Becker C, Ruschendorf F, Naville D, Begeot M, Khoo B, Nurnberg P, Huebner A, Cheetham ME, Clark AJ (2005) Mutations in MRAP, encoding a new interacting partner of the ACTH receptor, cause familial glucocorticoid deficiency type 2. Nat Genet 37:166–170

Meunier JC, Mollereau C, Toll L, Suaudeau C, Moisand C, Alvinerie P, Butour JL, Guillemot C, Ferrara P, Monsarrat B, Mazarguil H, Vassart G, Parmentier M, Costentin J (1995) Isolation and structure of the endogenous agonist of opioid receptor-like ORL1 receptor. Nature 377:532–535

Middleton J, Patterson AM, Gardner L, Schmutz C, Ashton BA (2002) Leukocyte extravasation: chemokine transport and presentation by the endothelium. Blood 100:3853–3860

Migeotte I, Riboldi E, Franssen JD, Gregoire F, Loison C, Wittamer V, Detheux M, Robberecht P, Costagliola S, Vassart G, Sozzani S, Parmentier M, Communi D (2005) Identification and characterization of an endogenous chemotactic ligand specific for FPRL2. J Exp Med 201:83–93

Mollereau C, Parmentier M, Mailleux P, Butour JL, Moisand C, Chalon P, Caput D, Vassart G, Meunier JC (1994) ORL1, a novel member of the opioid receptor family. Cloning, functional expression and localization. FEBS Lett 341:33–38

Mollereau C, Simons MJ, Soularue P, Liners F, Vassart G, Meunier JC, Parmentier M (1996) Structure, tissue distribution, and chromosomal localization of the prepronociceptin gene. Proc Natl Acad Sci USA 93:8666–8670

Mori M, Sugo T, Abe M, Shimomura Y, Kurihara M, Kitada C, Kikuchi K, Shintani Y, Kurokawa T, Onda H, Nishimura O, Fujino M (1999) Urotensin II is the endogenous ligand of a G-protein-coupled orphan receptor, SENR (GPR14). Biochem Biophys Res Commun 265:123–129

Murphy PM, Baggiolini M, Charo IF, Hebert CA, Horuk R, Matsushima K, Miller LH, Oppenheim JJ, Power CA (2000) International union of pharmacology. XXII. Nomenclature for chemokine receptors. Pharmacol Rev 52:145–176

Nagpal S, Patel S, Jacobe H, DiSepio D, Ghosn C, Malhotra M, Teng M, Duvic M, Chandraratna RA (1997) Tazarotene-induced gene 2 (TIG2), a novel retinoid-responsive gene in skin. J Invest Dermatol 109:91–95

Nathans J, Hogness DS (1984) Isolation and nucleotide sequence of the gene encoding human rhodopsin. Proc Natl Acad Sci USA 81:4851–4855

Nelson G, Hoon MA, Chandrashekar J, Zhang YF, Ryba NJP, Zuker CS (2001) Mammalian sweet taste receptors. Cell 106:381–390

Nelson G, Chandrashekar J, Hoon MA, Feng LX, Zhao G, Ryba NJ, Zuker CS (2002) An amino-acid taste receptor. Nature 416:199–202

Offermanns S, Simon MI (1995) G alpha 15 and G alpha 16 couple a wide variety of receptors to phospholipase C. J Biol Chem 270:15175–15180

Ohtaki T, Shintani Y, Honda S, Matsumoto H, Hori A, Kanehashi K, Terao Y, Kumano S, Takatsu Y, Masuda Y, Ishibashi Y, Watanabe T, Asada M, Yamada T, Suenaga M, Kitada C, Usuki S, Kurokawa T, Onda H, Nishimura O, Fujino M (2001) Metastasis suppressor gene KiSS-1 encodes peptide ligand of a G-protein-coupled receptor. Nature 411:613–617

Okuda-Ashitaka E, Minami T, Tachibana S, Yoshihara Y, Nishiuchi Y, Kimura T, Ito S (1998) Nocistatin, a peptide that blocks nociceptin action in pain transmission. Nature 392:286–289

Parameswaran N, Spielman WS (2006) RAMPs: the past, present and future. Trends Biochem Sci 31:631–638

Pin JP, Galvez T, Prezeau L (2003) Evolution, structure, and activation mechanism of family 3/C G-protein-coupled receptors. Pharmacol Ther 98:325–354

Reinscheid RK (2006) The orphanin FQ/nociceptin receptor as a novel drug target in psychiatric disorders. CNS Neurol Disord Drug Targets 5:219–224

Reinscheid RK, Nothaker HP, Bourson A, Ardati A, Henningsen R, Bunzow JR, Grandy DK, Langen H, Monsma FJ, Civelli O (1995) Orphanin FQ: a neuropeptide that activates an opioid-like G protein-coupled receptor. Science 270:792–794

Reisine T, Bell GI (1993) Molecular biology of opioid receptors. Trends Neurosci 16:506–510

Ribeiro S, Horuk R (2005) The clinical potential of chemokine receptor antagonists. Pharmacol Ther 107:44–58

Rossi D, Zlotnik A (2000) The biology of chemokines and their receptors. Ann Rev Immunol 18:217–242

Saito H, Kubota M, Roberts RW, Chi Q, Matsunami H (2004) RTP family members induce functional expression of mammalian odorant receptors. Cell 119:679–691

Saito Y, Nothacker HP, Wang Z, Lin SH, Leslie F, Civelli O (1999) Molecular characterization of the melanin-concentrating-hormone receptor. Nature 400:265–269

Sakurai T, Amemiya A, Ishii M, Matsuzaki I, Chemelli RM, Tanaka H, Williams SC, Richardson JA, Kozlowski GP, Wilson S, Arch JR, Buckingham RE, Haynes AC, Carr SA, Annan RS, McNulty DE, Liu WS, Terrett JA, Elshourbagy NA, Bergsma DJ, Yanagisawa M (1998) Orexins and orexin receptors: a family of hypothalamic neuropeptides and G protein-coupled receptors that regulate feeding behavior. Cell 92:573–585

Sallusto F, Mackay CR, Lanzavecchia A (2000) The role of chemokine receptors in primary, effector and memory immune responses. Ann Rev Immunol 18:593–620

Samson M, Edinger AL, Stordeur P, Rucker J, Verhasselt V, Sharron M, Govaerts C, Mollereau C, Vassart G, Doms RW, Parmentier M (1998) ChemR23, a putative chemoattractant receptor, is expressed in dendritic cells and is a coreceptor for SIV and some HIV-1 strains. Eur J Immunol 28:1689–1700

Shimomura Y, Harada M, Goto M, Sugo T, Matsumoto Y, Abe M, Watanabe T, Asami T, Kitada C, Mori M, Onda H, Fujino M (2002) Identification of neuropeptide W as the endogenous ligand for orphan G-protein-coupled receptors GPR7 and GPR8. J Biol Chem 277:35826–35832

Springael JY, Urizar E, Parmentier M (2005) Dimerization of chemokine receptors and its functional consequences. Cytokine Growth Factor Rev 16:611623

Springael JY, Nguyen PL, Urizar E, Costagliola S, Vassart G, Parmentier M (2006) Allosteric modulation of binding properties between units of chemokine receptor homo- and hetero-oligomers. Mol Pharmacol 69:1652–1661

Stables J, Green A, Marshall F, Fraser N, Knight E, Sautel M, Milligan G, Lee M, Rees S (1997) A bioluminescent assay for agonist activity at potentially any G-protein-coupled receptor. Anal Biochem 252:115–126

Takayasu S, Sakurai T, Iwasaki S, Teranishi H, Yamanaka A, Williams SC, Iguchi H, Kawasawa YI, Ikeda Y, Sakakibara I, Ohno K, Ioka RX, Murakami S, Dohmae N, Xie J, Suda T, Motoike T, Ohuchi T, Yanagisawa M, Sakai J (2006) A neuropeptide ligand of the G protein-coupled receptor GPR103 regulates feeding, behavioral arousal, and blood pressure in mice. Proc Natl Acad Sci USA 103:7438–7443

Taketani S, Adachi Y, Kohno H, Ikehara S, Tokunaga R, Ishii T (1998) Molecular characterization of a newly identified heme-binding protein induced during differentiation of urine erythroleukemia cells. J Biol Chem 273:31388–1394

Tanaka H, Yoshida T, Miyamoto N, Motoike T, Kurosu H, Shibata K, Yamanaka A, Williams SC, Richardson JA, Tsujino N, Garry MG, Lerner MR, King DS, O'Dowd BF, Sakurai T, Yanagisawa M (2003) Characterization of a family of endogenous neuropeptide ligands for the G protein-coupled receptors GPR7 and GPR. Proc Natl Acad Sci USA 100:6251–6256

Tatemoto K, Hosoya M, Habata Y, Fujii R, Kakegawa T, Zou MX, Kawamata Y, Fukusumi S, Hinuma S, Kitada C, Kurokawa T, Onda H, Fujino M (1998) Isolation and characterization of a novel endogenous peptide ligand for the human APJ receptor. Biochem Biophys Res Commun 251:471–476

Vermi W, Riboldi E, Wittamer V, Gentili F, Luini W, Marrelli S, Vecchi A, Franssen JD, Communi D, Massardi L, Sironi M, Mantovani A, Parmentier M, Facchetti F, Sozzani S (2005) Role of ChemR23 in directing the migration of myeloid and plasmacytoid dendritic cells to lymphoid organs and inflamed skin. J Exp Med 201:509–515

Wise A, Jupe SC, Rees S (2004) The identification of ligands at orphan G-protein coupled receptors. Annu Rev Pharmacol Toxicol 44:43–66

Wittamer V, Franssen JD, Vulcano M, Mirjolet JF, Le Poul E, Migeotte I, Brezillon S, Tyldesley R, Blanpain C, Detheux M, Mantovani A, Sozzani S, Vassart G, Parmentier M, Communi D (2003) Specific recruitment of antigen-presenting cells by chemerin, a novel processed ligand from human inflammatory fluids. J Exp Med 198:977–985

Wittamer V, Gregoire F, Robberecht P, Vassart G, Communi D, Parmentier M (2004) The C-terminal nonapeptide of mature chemerin activates the chemerin receptor with low nanomolar potency. J Biol Chem 279:9956–9962

Wittamer V, Bondue B, Guillabert A, Vassart G, Parmentier M, Communi D (2005) Neutrophil-mediated maturation of chemerin: a link between innate and adaptive immunity. J Immunol 175:487–493

Yarden Y, Escobedo JA, Kuang WJ, Yang-Feng TL, Daniel TO, Tremble PM, Chen EY, Ando ME, Harkins RN, Francke U, Fried VA, Ullrich A, Williams LT (1986) Structure of the receptor for platelet-derived growth factor helps define a family of closely related growth factor receptors. Nature 323:226–232

Xu YL, Reinscheid RK, Huitron-Resendiz S, Clark SD, Wang Z, Lin SH, Brucher FA, Zeng J, Ly NK, Henriksen SJ, de Lecea L, Civelli O (2004) Neuropeptide S: a neuropeptide promoting arousal and anxiolytic-like effects. Neuron 43:487–497

Zabel BA, Allen SJ, Kulig P, Allen JA, Cichy J, Handel TM, Butcher EC (2005) Chemerin activation by serine proteases of the coagulation, fibrinolytic, and inflammatory cascades. J Biol Chem 280:34661–34666

Zaveri N, Jiang F, Olsen C, Polgar W, Toll L (2005) Small-molecule agonists and antagonists of the opioid receptor-like receptor (ORL1, NOP): ligand-based analysis of structural factors influencing intrinsic activity at NOP. AAPS 7:e345–e352

Ernst Schering Foundation Symposium Proceedings, Vol. 2, pp. 187–209
DOI 10.1007/2789_2006_009

Published Online: 16 May 2007

Virus-Encoded G-Protein-Coupled Receptors: Constitutively Active (Dys)Regulators of Cell Function and Their Potential as Drug Target

H.F. Vischer, J.W. Hulshof, I.J.P. de Esch, M.J. Smit, R. Leurs(✉)

Leiden/Amsterdam Center for Drug Research (LACDR), Division of Medicinal Chemistry, Faculty of Sciences, Vrije Universiteit Amsterdam, De Boelelaan 1083, 1081 HV Amsterdam, The Netherlands
email: *r.leurs@few.vu.nl*

Abstract. G-protein-coupled receptors encoded by herpesviruses such as EBV, HCMV and KSHV are very interesting illustrations of the (patho)physiological importance of constitutive GPCR activity. These viral proteins are expressed on the cell surface of infected cells and often constitutively activate a variety of G-proteins. For some virus-encoded GPCRs, the constitutive activity has been shown to occur in vivo, i.e., in infected cells. In this paper, we will review the occurrence of virus-encoded GPCRs and describe their known signaling properties. Moreover, we will also review the efforts, directed towards the discovery of small molecule antagonist, that so far have been mainly focused on the HCMV-encoded GPCR US28. This virus-encoded receptor might be involved in cardiovascular diseases and cancer and seems an interesting target for drug intervention.

1 Introduction

The chemokines and their receptors play a key role in the regulation of the immune system and other important pathophysiological conditions, such as organogenesis, angiogenesis, metastasis, and growth of tumor cells (Murphy et al. 2000). The mammalian chemokine system (e.g., human, mouse, and rat) is made up of approximately 45 small chemokine ligands and 20 chemokine receptors (Murphy 2002). Chemokines are a family of small proteins that adopt a similar tertiary folding, even in cases of low overall sequence identity (varying from 20% to 95%). The various protein ligands interact with selected chemokine receptors, which belong to the membrane-associated G-protein-coupled receptor (GPCR) family. Chemokine receptors are classified (i.e., CCR1-11, CXCR1-6, CX3CR1, and XCR1) according to their ability to bind a specific subclass of chemokines (Murphy 2002). Given the prominent role of chemokine receptors in regulating intracellular signaling in response to chemokine ligands, these receptors are very promising targets for immunomodulatory therapy (Onuffer and Horuk 2002; Gao and Metz 2003). Interestingly, such receptors are also employed by several viruses in order to subvert the immune system and/or redirect intracellular signaling for their own benefit (Alcami 2003; Vischer et al. 2006a, 2006b). Perhaps the best-known employment of GPCRs by viruses is the use of various mammalian chemokine receptors as HIV entry factors (Ray and Doms 2006). However, a variety of herpesviruses possess genes encoding proteins with homology to human GPCRs. The virus-encoded GPCRs are expressed in infected cells after viral infection (Fig. 1) and are currently considered important viral proteins that might be essential for the viral re-routing of cellular function.

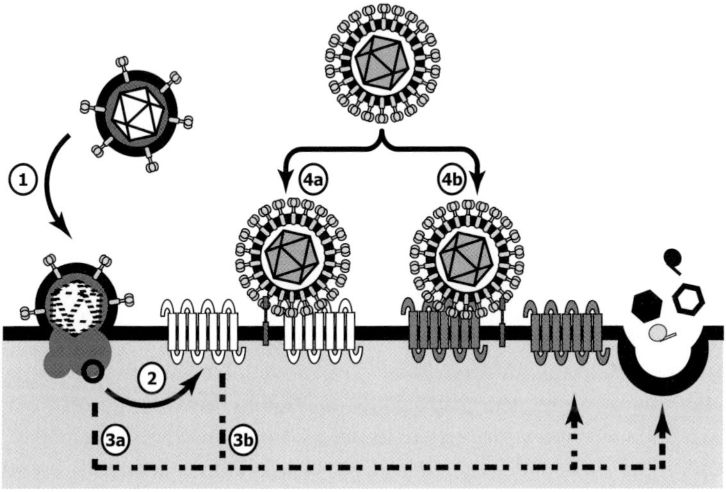

Fig. 1. Dysregulation of cell functioning by virally encoded GPCRs: herpesviruses may penetrate host cells by membrane fusion of the viral envelope with the host cell membrane (*1*). The viral nucleocapsid is rapidly translocated to the nucleus upon entry into the cytoplasm and viral gene expression is subsequently initiated (*2*). Virally encoded GPCRs (*white*) are expressed on the cell membrane and may modulate cell functioning in response to natural ligands and/or constitutively. Virus-encoded GPCRs (*3b*) as well as other virus-encoded proteins (*3a*) may modulate the expression and/or secretion of host receptors (*gray*) and/or autocrine/paracrine factors, respectively. In addition, the HCMV-encoded GPCR US28 (*4a*) functions as an additional coreceptor for CD4-mediated HIV entry next to the cellular chemokine receptors CCR5 and CXCR4 (*4b*)

2 Virus-Encoded G-Protein-Coupled Receptors

Genes encoding homologs of cellular GPCRs have been identified in a number of herpesviruses and poxviruses. These double-stranded DNA viruses have presumably acquired such genes from the host genome by retrotransposition during the course of their intimate co-evolution (Davison et al. 2002). Interestingly, most virus-encoded GPCRs display highest sequence identity to the subfamily of chemotactic cytokine (chemokine) GPCRs. Some viral mimics of host chemokine receptors are indeed responsive to chemokines and may as such serve viral spread by immune evasion, chemotaxis of infected cells and prime host cells for productive viral replication. Other virally-encoded GPCRs (vGPCR) are still orphan, as no natural ligand has been identified yet. In contrast to cellular chemokine receptors, a number of virally-encoded chemokine receptors (vCR) activates intracellular signal transduction networks in a ligand-independent manner (Sodhi et al. 2004a; Vischer et al. 2006b).

3 Herpesvirus-Encoded GPCRs

Herpesviruses are characterized by their ability to establish lifelong latent and persistent infection. Latent viral infection is usually without severe clinical consequences in healthy individuals. However, reactivation of productive viral replication in immunocompromised patients may result in profound acute and chronic life-threatening pathologies. Based on their genomic structure, herpesviruses are classified into three subfamilies: the α-, β,-, and γ-herpesviruses. At least one GPCR-encoding gene is present in the genome of β- and γ-herpesviruses, whereas α-herpesviruses lack such genes (Vischer et al. 2006b).

3.1 Roseoloviruses and Cytomegaloviruses

The subfamily of β-herpesviruses consists of roseoloviruses and cytomegaloviruses, which are ubiquitously spread in the general population. Three highly related species of roseolovirus have been isolated from human: human herpesvirus (HHV) 6A, HHV-6B, and HHV-7. Primary infection occurs usually during early childhood, and causes exanthem subitum (roseola). Roseoloviruses persist latently in monocytes but replicates most efficiently in $CD4^+$ T lymphocytes. Roseolovirus reactivation in immunocompromised individuals (post-transplantation or HIV patients) may result in encephalitis, pneumonitis, hepatitis, graft rejection and bone marrow suppression (De Bolle et al. 2005). Roseoloviruses encode two GPCR homologs, namely U12 and U51. Both GPCRs display sequence identity to human chemokine receptors (<20%) and are highly responsive to a variety of CC chemokines. The inflammatory chemokines CCL2, CCL3, CCL4, and CCL5 induced intracellular Ca^{2+} mobilization in HHV-6-encoded U12-expressing cells

(Isegawa et al. 1998). In contrast, its HHV-7-encoded ortholog was not responsive to CCL2 and CCL5, but mediated Ca^{2+} signaling upon stimulation with CCL17, CCL19, CCL21, and CCL22 (Nakano et al. 2003; Tadagaki et al. 2005). Interestingly, the CCR7 ligands CCL19 and CCL21 were able to induce HHV-7 U12-mediated chemotaxis of Jurkat cells, allowing homing of HHV-7 infected cells into lymph nodes, which may contribute to viral dissemination (Tadagaki et al. 2005).

The U51 receptor of HHV-6 efficiently binds inflammatory chemokines CCL2, CCL5, CCL7, CCL11, CCL13, as well as the HHV-8-encoded viral macrophage inflammatory protein II (Milne et al. 2000). Expression of this receptor in COS-7 cells results in the constitutive activation of phospholipase C and inhibition cAMP-responsive-element (CRE)-mediated gene transcription by coupling through $G_{q/11}$ proteins (Fitzsimons et al. 2006). Interestingly, CCL2, CCL5, and CCL11 differently modulate constitutive signaling of U51 by directing receptor coupling to distinct G-proteins. U51-mediated signaling downregulates transcription of the CCL5 gene, as such contributing to immune modulation during viral infection (Milne et al. 2000). The U51 ortholog encoded by HHV-7 induced intracellular Ca^{2+} mobilization in response to the same chemokine subset as HHV-7 U12, however, these chemokines were not able to induce U51-mediated chemotaxis of Jurkat cells (Tadagaki et al. 2005). Infection of permissive T lymphocytes with HHV-6B revealed that U12 is a late gene, whereas U51 is an early gene (Oster and Hollsberg 2002).

Cytomegalovirus (CMV) infects endothelial, epithelial, and smooth muscle cells of the upper gastrointestinal, respiratory, and urogenital tract (Landolfo et al. 2003) and is spread throughout the body by latently infected monocytes (Streblow and Nelson 2003). Differentiation of these monocytes into macrophages is accompanied by reactivation of CMV, leading to the release of infectious virions (Streblow and Nelson 2003). CMV infection or reactivation in an immunocompromised host may result in severe complications such as damage of liver, brain, retina, and lung (Landolfo et al. 2003). In addition, increasing evidence indicates that CMV may also contribute to inflammatory and autoimmune diseases as well as cancer (Soderberg-Naucler 2006). The human CMV (HCMV) genome contains four GPCR-encoding genes, namely UL33, UL78, US27, and US28 (Dolan et al. 2004). The UL33 and UL78 genes are conserved in genomes of all sequenced cytomegaloviruses and correspond, respectively, to the U12 and U51 genes of roseoloviruses with respect to gene positioning and orientation. An additional GPCR-encoding gene cluster is present on the unique short (US) region of the genome of primate CMVs, consisting of the adjacent genes US27 and US28 in HCMV and chimpanzee (C)CMV, and five juxtaposed genes in rhesus macaque (Rh)CMV and African green monkey (S)CMV (Davison et al. 2003; Hansen et al. 2003; Sahagun-Ruiz et al. 2004).

The UL33 genes display highest amino acid sequence identity to CCR10 and CCR3. Nevertheless, HCMV-encoded UL33 and its rat CMV (RCMV)-encoded ortholog R33 are orphan receptors displaying neither responsiveness nor affinity for chemokines (Gruijthuijsen et al. 2002; Casarosa et al. 2003a). In contrast, mouse CMV (MCMV)-encoded M33-mediated mCCL5-directed chemotaxis of vascular smooth muscle cells (Melnychuk et al. 2005). UL33 orthologs of HCMV, RCMV, and MCMV all constitutively activate multiple signaling pathways through various G-proteins (Gruijthuijsen et al. 2002; Waldhoer et al. 2002; Casarosa et al. 2003a). In contrast to U51 of roseoloviruses, the CMV-encoded UL78 is not responsive to chemokines and has not been shown to modulate intracellular signaling pathways. The US27 and US28 proteins are homologs of CXCR3 (i.e., 23% sequence identity) and CX3CR1 (i.e., 36% sequence identity), respectively (Vischer et al. 2006b). Hitherto, US27 is still an orphan receptor and appeared not to effect intracellular signaling (Waldhoer et al. 2002). In contrast, US28 is able to bind a wide variety of CC chemokines as well as CX3CL1 and modulated intracellular signaling pathways both constitutively and upon chemokine activation (Couty and Gershengorn 2005; Vischer et al. 2006a). In addition, US28 is constitutively phosphorylated, which is followed by adaptin and dynamin-dependent internalization via clathrin-coated pits (Mokros et al. 2002; Fraile-Ramos et al. 2003; Droese et al. 2004).

US28 is transcribed in latently infected monocytes, allowing immune evasion and/or US28-mediated chemotaxis of these cells along chemokine gradients as well as tethering to membrane-bound CX3CL1 expressed on vascular endothelial cells (Beisser et al. 2001). During productive viral infection in permissive cells, the US28 and UL78 genes are expressed with early kinetics, requiring immediate-early-protein-mediated transcriptional activation, whereas US27 and UL33 genes are transcribed with late kinetics after the onset of viral DNA replication (Mocarski and Courcelle 2001). The UL33, UL78, US28, and presumably US27 receptor proteins are subsequently incorporated in the viral envelope (Margulies et al. 1996; Fraile-Ramos et al. 2001; Oliveira and Shenk 2001; Fraile-Ramos et al. 2002; Penfold et al. 2003). Expression of these proteins is not essential for viral infection of permissive cell in vitro (Margulies et al. 1996; Davis-Poynter et al. 1997; Bodaghi et al. 1998; Vieira et al. 1998; Beisser et al. 1999; Michel et al. 2005). However, deletion or disrupting R33/M33 prevented viral dissemination to the salivary glands (Davis-Poynter et al. 1997; Beisser et al. 1999), attenuated RCMV-accelerated transplant vascular sclerosis and chronic rejection (Streblow et al. 2005), and resulted in a lower mortality rate compared with wild-type RCMV-infected rats (Beisser et al. 1998). Although deletion of UL78 did not affect HCMV replication in vitro (Michel et al. 2005), disruption of its RCMV and MCMV orthologs attenuated viral replication in vitro and in vivo as compared with wild-type virus (Beisser et al. 1998; Oliveira and Shenk 2001; Kaptein et al. 2003).

3.2 HCMV-Encoded GPCRs as Drug Targets?

Both US28 and UL33 alter cellular signaling in a constitutively active manner when ectopically expressed and more importantly after HCMV infection, as shown using HCMV US28 and UL33 deletion strains (Streblow et al. 1999; Casarosa et al. 2001, 2003a; Minisini et al. 2003). Multiple signaling networks, including effectors and transcription factors, are constitutively activated within infected cells in part by these viral receptors, reprograming the cellular machinery to modulate cellular function after infection. These findings in fact indicate the relevance of constitutive receptor activity, which for GPCRs in general is often difficult to prove in vivo.

Moreover, both viral receptors show promiscuous G-protein-coupling (Waldhoer et al. 2002; Casarosa et al. 2001, 2003a; Minisini et al. 2003). The chemokine receptor homologs, on the other hand, do not or display only limited ligand-independent signaling and activate primarily Gi/o proteins (Offermanns 2003). The broad-spectrum binding profile of US28 suggests that US28 could act as a chemokine scavenger and thereby aid in subversion of the immune system (Kuhn et al. 1995; Kledal et al. 1998). In addition, US28-mediated constitutive signaling potentiates chemokine-induced signaling of the Gi-coupled CCR1 (Bakker et al. 2004). Since HCMV primarily infects leukocytes, smooth muscle and endothelial cells, in which chemokine receptors play prominent roles, HCMV-encoded receptor expression may alter ligand-induced signaling via these receptors and contribute to the CMV-induced pathology.

HCMV has been associated with chronic diseases, including, for example, vascular diseases (Stassen et al. 2006) and malignancies such as malignant glioma, colon, and prostate cancers (Cobbs et al. 2002; Harkins et al. 2002). Although the causative role for HCMV in the development of vascular disease and malignancies remains to be established, various HCMV proteins and DNA have been detected with high frequency in atherosclerotic plaques and tumor tissues (Cobbs et al. 2002; Harkins et al. 2002; Samanta et al. 2003). A molecular basis for the involvement of HCMV and US28 in the progression of atherosclerosis has been provided by the fact that infection of smooth muscle cells with CMV leads to an US28-dependent migration (Streblow et al. 1999). Moreover, the CMV-encoded receptors US28 and UL33 constitutively activate NF-κB, a transcription factor that plays a critical role in the regulation of inducible genes in immune response and inflammatory events associated with, for example, atherosclerosis (Chen et al. 1999).

HCMV infection also upregulates different growth factors and cytokines, resulting in enhanced cell survival, proliferation, and angiogenesis (Cinatl et al. 2004). We have recently observed that expression of US28 in vitro induces a transformed and pro-angiogenic phenotype (Maussang et al. 2006). Also in HCMV-infected cells, activation of pro-angiogenic signaling pathways was apparent and could in part be attributed to US28. As such, after HCMV infection US28 might act in a concerted manner with other HCMV-encoded proteins, which were previously linked to oncogenesis (Cinatl et al. 2004), and enhance and/or promote tumorigenesis.

Finally, US28 has been reported as a coreceptor for HIV infection. Similarly to the chemokine receptors CCR5 and CXCR4, US28 exhibits HIV cofactor activity when coexpressed with CD4 (Pleskoff et al. 1997).

To further prove a causative link between the HCMV-encoded receptors and chronic diseases, the development of adequate disease model systems that allow in vivo analyses of all four (H)CMV-encoded GPCR subtypes is essential. The availability of vGPCR-knockout strains of (H/C)CMV and specific (pharmacological or RNAi) inhibitors targeting these receptors is of importance also, to elucidate the contribution of HCMV-encoded GPCRs to HCMV pathogenesis and reveal their potential as future drug target.

3.3 Rhadinoviruses and Lymphocryptoviruses

The rhadinoviruses and lymphocryptoviruses constitute the family of γ-herpesviruses. The human rhadinovirus herpesvirus 8 (HHV-8), or Kaposi's sarcoma herpesvirus (KSHV), was first isolated from Kaposi's sarcoma skin lesions of an AIDS patient (Chang et al. 1994). KSHV displays a relatively low infectivity rate (<5%) in the general population as compared with other herpesviruses. KSHV persists latently in pre- and postgerminal center B lymphocytes and endothelial precursor cells (Dupin et al. 1999). Reactivation of KSHV in immunosuppressed individuals (e.g., AIDS patients) causes multifocal angioproliferative Kaposi's sarcoma lesions, multicentric Castleman's disease, and/or primary effusion lymphoma. KSHV encodes a single GPCR, namely ORF74, which displays highest sequence identity to CXCR2. ORF74 is predominantly expressed during the early lytic phase of viral replication, and is detected in only a fraction (1%–5%) of the tumor cells of Kaposi's sarcoma lesion biopsies (Kirshner et al. 1999; Sun et al. 1999). ORF74 constitutively activates multiple intracellular signaling pathways by coupling to $G_{q/11}$, $G_{i/o}$, and G_{13}, which can be modulated by various CXC chemokines (i.e., CXCL1, 8, 10, and 12) and KSHV-encoded vMIP-II (Rosenkilde et al. 1999; Sodhi et al. 2004a). Transgenic expression of ORF74 in mice induced Kaposi's sarcoma-like tumors. Continuous ORF74-mediated constitutive activation of Akt/protein kinase B is essential for the initiation and progression of Kaposi's sarcomagenesis in mouse models (Sodhi et al. 2004b; Jensen et al. 2005). Indeed, high phosphorylated (activated) levels of this antiapoptotic serine-threonine kinase have been observed in biopsies of Kaposi's sarcoma from human patients (Sodhi et al. 2004b). ORF74 activates Akt both directly via PI3K, phospholipase C-dependent protein kinase C, and p44/42 MAPK (Montaner et al. 2001; Smit et al. 2002), as well as indirectly by upregulating the expression of both VEGF and its cognate receptor KDR2 (Bais et al. 1998, 2003; Sodhi et al. 2000). In fact, ORF74-induced upregulation and release of pro-angiogenic factors, pro-inflammatory cytokines and chemokines from a few cells, is sufficient to drive angioproliferative tumor formation by autocrine stimulation (Grisotto et al. 2006) or paracrine stimulation of (latently-infected) neighboring cells (Montaner et al. 2003; Sodhi et al. 2004a; Jensen et al. 2005; Montaner et al. 2006). With the exception of bovine herpesvirus 4 and alcelaphine herpesvirus 1, all hitherto isolated rhadinoviruses contain an ORF74-encoding gene (McGeoch 2001). Interestingly, nonhuman rhadinoviruses display narrower G-protein coupling promiscuity (reviewed in Vischer et al. 2006b). Moreover, murine γ-herpesvirus 68-encoded ORF74 is devoid of constitutivity, despite its capacity to induce focus formation when expressed in NIH3T3 cells (Verzijl et al. 2004).

Hitherto, lymphocryptoviruses (~44 distinct species) have only been isolated from "higher" order primates (Ehlers et al. 2003). The human lymphocryptovirus HHV-4 or Epstein-Barr virus (EBV) is widely spread in the general population (>90%) and establishes life-long, persistent latent infections in memory B lymphocytes (Wang et al. 2001). To this end, EBV infects naïve B cells and drives proliferation and transition of these cells into memory B lymphocytes by means of a regulated cascade of viral transcription programs (Thorley-Lawson 2005). Infected B cells that are blocked in this transition and express EBV proteins (epitopes) are normally efficiently recognized and eliminated by cytotoxic T cells. However, immunocompromised patients (e.g., post-transplant immunosuppression, AIDS, and malaria) are at particular risk of developing EBV-associated (Hodgkin's and Burkitt's) lymphomas (Middeldorp et al. 2003; Thorley-Lawson and Gross 2004). In addition, EBV may also be causative to nasal natural killer (NK-) T cell lymphoma, nasopharyngeal and gastric carcinoma, oral hairy leukoplakia, and leiomyosarcoma (Thompson and Kurzrock 2004). Lymphocryptoviruses encode a single GPCR protein, namely BILF1, which displays very limited sequence identity (<15%) to host GPCRs (Vischer et al. 2006b). Hitherto, BILF1 is still an orphan receptor as a natural ligand for this receptor remains to be identified. Importantly, BILF1 constitutively modulates CRE- and NF-κB signaling pathways by coupling to Gi proteins (Beisser et al. 2005; Paulsen et al. 2005). Moreover, BILF1 may attenuate the cellular antiviral response by constitutively inhibiting the phosphorylation of RNA-dependent protein kinase (Beisser et al. 2005).

4 Poxvirus-Encoded GPCRs

Poxviruses are epitheliotropic and cause acute febrile illness accompanied by blistered skin lesions that form pockmarks. Poxviruses can be transmitted by direct contact or via the respiratory tract (Diven 2001). Poxviruses are often not restricted to a single host species and may therefore reside in a reservoir host causing mild, subclinical pathologies. However, transmission to a zoonotic host may cause severe pathologies (McFadden 2005). Infections are often self-limiting; however, some poxvirus species cause life-threatening infections in certain hosts. One or more GPCR-encoding genes are present in the genomes of avipoxviruses, capripoxviruses, suipoxviruses, and yatapoxvirus, whereas other poxvirus genera lack such genes (Vischer et al. 2006b). Avipoxvirus species contain three or four GPCR-encoding genes, which share amino acid sequence identity with CXC chemokine receptors. Suipoxviruses and capripoxviruses encode a single GPCR with highest sequence identity to CCR8, whereas yatapoxvirus species encode two CCR8 homologs. Indeed yaba-like disease virus-encoded 7L protein, but not 145R, display a similar chemokine-binding profile as CCR8 (Najarro et al. 2003, 2006). The 7L protein is expressed as an early as well as late gene during productive viral infection (Najarro et al. 2003, 2006).

5 Small Nonpeptidergic Ligands Acting on Viral-Encoded GPCRs

Currently, the HCMV-encoded receptor US28 is the only viral-encoded GPCR for which small non-peptide molecules have been identified.

Several classes of ligands that inhibit chemokine binding to US28 have been reported in the patent literature. Additionally, a series of inverse agonists acting on US28 were published recently (Casarosa et al. 2003b; Hulshof et al. 2005). In 2002, Chemocentryx Inc. disclosed that CMV dissemination during CMV infection in a host could be inhibited by compounds that reversibly block chemokine binding to US28 (Schall et al. 2002a, 2002b). Among these reported US28 ligands are a series of piperazinyldibenzothiepins, represented by the 5-HT receptor antagonist methiothepin (1) and the D_2 dopamine/5-HT2 antagonist octoclothepin (2) (Fig. 2). These tricyclic compounds have been shown to specifically inhibit ^{125}I-CX3CL1 binding to US28-expressing cells in a reversible manner with IC_{50} values of 0.3 μM and 0.7 μM, respectively (Schall et al. 2002a). In cytoplasmic calcium mobilization experiments in US28-expressing HEK293 cells, CX3CL1 acted as an agonist by inducing a rise in intracellular Ca^{2+}. Interestingly, both methiothepin and octoclothepin also acted as agonists in the same assay and were able to desensitize the subsequent Ca^{2+} mobilization by CX3CL1.

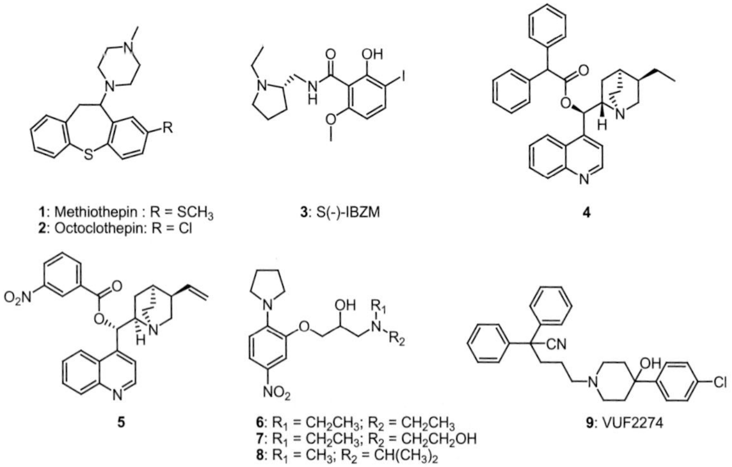

Fig. 2. Chemical structures of ligands acting on the HCMV encoded GPCR US28

A family of benzamides, exemplified by *S*(-)-IBZM (3) in Fig. 2, have been reported as ligands interacting with US28 (Schall et al. 2002b). *S*(-)-IBZM is known as a D_2 dopamine receptor ligand, which can be radiolabeled due to the presence of an accessible iodide substituent in the structure. Radiolabeled 3 is used in the clinic to visualize D_2-dopamine receptors in the human brain in vivo by single photon emission tomography (SPECT). (Laruelle and Abi-Dargham 1999). As shown for methiothepin and octoclothepin, *S*(-)-IBZM was able to specifically displace ^{125}I-CX3CL1 binding to US28 with an IC_{50} value of 0.6 μM and to act as an agonist in the calcium mobilization assay. Due to the observed interaction with US28, it was claimed that ^{123}I-IBZM could be used for the in vivo detection, diagnosis, and imaging of CMV infection in a host using SPECT (or PET if ^{18}F or another positron emitter is used). Yet, in view of the relatively low micromolar affinity of *S*(-)-IBZM, the feasibility of this approach seems questionable.

A large series of bicyclic compounds, represented by compounds 4 and 5 (Fig. 2), were reported for the treatment or prevention of viral dissemination from CMV infection by inhibiting chemokine binding to US28 (McMaster et al. 2003b). These cinchonine or cinchonidine derivatives were evaluated for their binding properties by their ability to inhibit ^{125}I-CX3CL1 binding to human US28-expressing cells and to rhesus CMV-infected rhesus dermal fibroblasts. Interestingly, a striking species selectivity between human and rhesus US28 was demonstrated. Several tested ligands exhibited IC_{50} values less than 1 μM on human US28 or rhesus US28, but only compound 4 showed a binding affinity less than 1 μM on both vGPCRs. Recently, species selectivity has been shown for CCR1 antagonists as well and this could be problematic if these molecules are tested in animal models of disease (Liang et al. 2000).

The last group of compounds that were disclosed by Chemocentryx Inc. as inhibitors of chemokine binding to US28 is exemplified by the compounds 6–8 (Fig. 2) (McMaster et al. 2003a). The activities of these arylamines were determined by their ability to inhibit ^{125}I-CX3CL1 binding to US28-expressing cells. As for the bicyclic compounds, the most active compounds were claimed to have IC_{50} values under 1 μM; however, specific IC_{50} values and functional data were not reported.

Recently, the synthesis and structure–activity relationships of a whole series of 4-substituted piperidine derivatives, represented by lead compound VUF2274 (9), were reported (Fig. 2) (Casarosa et al. 2003b; Hulshof et al. 2005). VUF2274 has been shown to inhibit ^{125}I-CCL5 binding to US28-expressing COS-7 cells with an IC_{50} value of 9.3 μM. In previous studies, we have shown that in transiently transfected COS-7 cells as well as in HCMV-infected cells, US28 constitutively activates phospholipase C and the transcription factor NF-κB in an agonist-independent manner (Casarosa et al. 2001, 2003b). Constitutive signaling of US28 could be completely blocked by VUF2274 with an EC_{50} value of 3.2 μM. This molecule has been previously reported as an antagonist on the human chemokine receptor CCR1 (Hesselgesser et al. 1998) and was screened on US28 because of the sequence homology of this viral receptor with CCR1 (33% identity). VUF2274 does not only dose-dependently inhibit the US28-mediated constitutive activation of PLC in both transiently transfected cells and HCMV-infected fibroblasts, but also inhibits 60% of the US28-mediated HIV entry in cells co-transfected with CD4 (Casarosa et al. 2003b). As expected, it was shown that VUF2274 and the relatively large chemokines CX3CL1 and CCL5 do not share the same binding site on US28, so the small molecule acts as an allosteric modulator on US28 (Casarosa et al. 2003b). SAR studies on the VUF2274 scaffold revealed that a 4-phenylpiperidine moiety is essential for affinity and activity (Hulshof et al. 2005). Other structural changes of 9 were shown to be less important. Currently, more extensive SAR studies are ongoing to improve the affinity and potency for US28 as well as a better selectivity for this receptor (Fig. 2). These molecules are being used as tools to further investigate the significance of constitutive signaling of US28 and its influence in viral infection.

At present, no low-molecular-weight ligands are known to act on other vGPCRs, but this could be only a matter of time. Previously, it was shown that ORF74, a constitutively active GPCR encoded by KSHV, was susceptible to non-peptide inverse agonists by inhibition of the constitutive signaling of the viral receptor by a Zn^{2+} ion (Rosenkilde et al. 1999). To this end, a silent metal ion site was constructed by His-substitution of Arg^{208} and Arg^{212} and the signaling of this ORF74 mutant receptor was blocked by Zn^{2+}, which was acting as an inverse agonist with an EC_{50} around 1 μM. This was suggested to be a proof of concept that it should be possible to identify small nonpeptidergic inverse agonists targeted toward the extracellular part of this viral-encoded GPCR.

6 Conclusion

In summary, for many of the virus-encoded GPCRs, information on their pharmacological properties is becoming available. The viral proteins often act in a constitutive manner and can be seen as versatile signaling devices, which are essential to drive the viral re-routing of cellular signaling. The search for small ligands acting on vGPCRs is still an unexplored but fascinating research field. The identification of these non-peptidergic molecules is essential to investigate the role of vGPCRs in the pathogenesis of viral infection and these molecules can be considered promising therapeutics for clinical antiviral intervention.

References

Alcami A (2003) Viral mimicry of cytokines, chemokines and their receptors. Nat Rev Immunol 3:36–50

Bais C, Santomasso B, Coso O, Arvanitakis L, Raaka EG, Gutkind JS, Asch AS, Cesarman E, Gershengorn MC, Mesri EA, Gerhengorn MC (1998) G-protein-coupled receptor of Kaposi's sarcoma-associated herpesvirus is a viral oncogene and angiogenesis activator. Nature 391:86–89

Bais C, Van Geelen A, Eroles P, Mutlu A, Chiozzini C, Dias S, Silverstein RL, Rafii S, Mesri EA (2003) Kaposi's sarcoma associated herpesvirus G protein-coupled receptor immortalizes human endothelial cells by activation of the VEGF receptor-2/KDR. Cancer Cell 3:131–143

Bakker RA, Casarosa P, Timmerman H, Smit MJ, Leurs R (2004) Constitutively active Gq/11-coupled Receptors Enable Signaling by Co-expressed Gi/o-coupled Receptors. J Biol Chem 279:5152–5161

Beisser PS, Vink C, Van Dam JG, Grauls G, Vanherle SJ, Bruggeman CA (1998) The R33 G protein-coupled receptor gene of rat cytomegalovirus plays an essential role in the pathogenesis of viral infection. J Virol 72:2352–2363

Beisser PS, Grauls G, Bruggeman CA, Vink C (1999) Deletion of the R78 G protein-coupled receptor gene from rat cytomegalovirus results in an attenuated, syncytium-inducing mutant strain. J Virol 73:7218–7230

Beisser PS, Laurent L, Virelizier JL, Michelson S (2001) Human cytomegalovirus chemokine receptor gene US28 is transcribed in latently infected THP-1 monocytes. J Virol 75:5949–5957

Beisser PS, Verzijl D, Gruijthuijsen YK, Beuken E, Smit MJ, Leurs R, Bruggeman CA, Vink C (2005) The Epstein–Barr Virus BILF1 Gene Encodes a G Protein-Coupled Receptor That Inhibits Phosphorylation of RNA-Dependent Protein Kinase. J Virol 79:441–449

Bodaghi B, Jones TR, Zipeto D, Vita C, Sun L, Laurent L, Arenzana-Seisdedos F, Virelizier JL, Michelson S (1998) Chemokine sequestration by viral chemoreceptors as a novel viral escape strategy: withdrawal of chemokines from the environment of cytomegalovirus-infected cells. J Exp Med 188:855–866

Casarosa P, Bakker RA, Verzijl D, Navis M, Timmerman H, Leurs R, Smit MJ (2001) Constitutive signaling of the human cytomegalovirus-encoded chemokine receptor US28. J Biol Chem 276:1133–1137

Casarosa P, Gruijthuijsen YK, Michel D, Beisser PS, Holl J, Fitzsimons CP, Verzijl D, Bruggeman CA, Mertens T, Leurs R, Vink C, Smit MJ (2003a) Constitutive signaling of the human cytomegalovirus-encoded receptor UL33 differs from that of its rat cytomegalovirus homolog R33 by promiscuous activation of G proteins of the Gq, Gi, and Gs classes. J Biol Chem 278:50010–50023

Casarosa P, Menge WM, Minisini R, Otto C, van Heteren J, Jongejan A, Timmerman H, Moepps B, Kirchhoff F, Mertens T, Smit MJ, Leurs R (2003b) Identification of the first nonpeptidergic inverse agonist for a constitutively active viral-encoded G protein-coupled receptor. J Biol Chem 278:5172–5178

Chang Y, Cesarman E, Pessin MS, Lee F, Culpepper J, Knowles DM, Moore PS (1994) Identification of herpesvirus-like DNA sequences in AIDS-associated Kaposi's sarcoma. Science 266:1865–1869

Chen F, Castranova V, Shi X, Demers LM (1999) New insights into the role of nuclear factor-kappaB, a ubiquitous transcription factor in the initiation of diseases. Clin Chem 45:7–17

Cinatl J Jr, Vogel JU, Kotchetkov R, Wilhelm Doerr H (2004) Oncomodulatory signals by regulatory proteins encoded by human cytomegalovirus: a novel role for viral infection in tumor progression. FEMS Microbiol Rev 28:59–77

Cobbs CS, Harkins L, Samanta M, Gillespie GY, Bharara S, King PH, Nabors LB, Cobbs CG, Britt WJ (2002) Human cytomegalovirus infection and expression in human malignant glioma. Cancer Res 62:3347–3350

Couty JP, Gershengorn MC (2005) G-protein-coupled receptors encoded by human herpesviruses. Trends Pharmacol Sci 26:405–411

Davison AJ, Dargan DJ, Stow ND (2002) Fundamental and accessory systems in herpesviruses. Antiviral Res 56:1–11

Davison AJ, Dolan A, Akter P, Addison C, Dargan DJ, Alcendor DJ, McGeoch DJ, Hayward GS (2003) The human cytomegalovirus genome revisited: comparison with the chimpanzee cytomegalovirus genome. J Gen Virol 84:17–28

Davis-Poynter NJ, Lynch DM, Vally H, Shellam GR, Rawlinson WD, Barrell BG, Farrell HE (1997) Identification and characterization of a G protein-coupled receptor homolog encoded by murine cytomegalovirus. J Virol 71:1521–1529

De Bolle L, Naesens L, De Clercq E (2005) Update on human herpesvirus 6 biology, clinical features, and therapy. Clin Microbiol Rev 18:217–245

Diven DG (2001) An overview of poxviruses. J Am Acad Dermatol 44:1–16

Dolan A, Cunningham C, Hector RD, Hassan-Walker AF, Lee L, Addison C, Dargan DJ, McGeoch DJ, Gatherer D, Emery VC, Griffiths PD, Sinzger C, McSharry BP, Wilkinson GW, Davison AJ (2004) Genetic content of wild-type human cytomegalovirus. J Gen Virol 85:1301–1312

Droese J, Mokros T, Hermosilla R, Schulein R, Lipp M, Hopken UE, Rehm A (2004) HCMV-encoded chemokine receptor US28 employs multiple routes for internalization. Biochem Biophys Res Commun 322:42–49

Dupin N, Fisher C, Kellam P, Ariad S, Tulliez M, Franck N, van Marck E, Salmon D, Gorin I, Escande JP, Weiss RA, Alitalo K, Boshoff C (1999) Distribution of human herpesvirus-8 latently infected cells in Kaposi's sarcoma, multicentric Castleman's disease, and primary effusion lymphoma. Proc Natl Acad Sci USA 96:4546–4551

Ehlers B, Ochs A, Leendertz F, Goltz M, Boesch C, Matz-Rensing K (2003) Novel simian homologues of Epstein-Barr virus. J Virol 77:10695–10699

Fitzsimons CP, Gompels UA, Verzijl D, Vischer HF, Mattick C, Leurs R, Smit MJ (2006) Chemokine-directed trafficking of receptor stimulus to different g proteins: selective inducible and constitutive signaling by human herpesvirus 6-encoded chemokine receptor U51. Mol Pharmacol 69:888–898

Fraile-Ramos A, Kledal TN, Pelchen-Matthews A, Bowers K, Schwartz TW, Marsh M (2001) The human cytomegalovirus US28 protein is located in endocytic vesicles and undergoes constitutive endocytosis and recycling. Mol Biol Cell 12:1737–1749

Fraile-Ramos A, Pelchen-Matthews A, Kledal TN, Browne H, Schwartz TW, Marsh M (2002) Localization of HCMV UL33 and US27 in endocytic compartments and viral membranes. Traffic 3:218–232

Fraile-Ramos A, Kohout TA, Waldhoer M, Marsh M (2003) Endocytosis of the viral chemokine receptor US28 does not require beta-arrestins but is dependent on the clathrin-mediated pathway. Traffic 4:243–253

Gao Z, Metz WA (2003) Unraveling the chemistry of chemokine receptor ligands. Chem Rev 103:3733–3752

Grisotto MG, Garin A, Martin AP, Jensen KK, Chan P, Sealfon SC, Lira SA (2006) The human herpesvirus 8 chemokine receptor vGPCR triggers autonomous proliferation of endothelial cells. J Clin Invest 116:1264–1273

Gruijthuijsen YK, Casarosa P, Kaptein SJ, Broers JL, Leurs R, Bruggeman CA, Smit MJ, Vink C (2002) The rat cytomegalovirus R33-encoded G protein-coupled receptor signals in a constitutive fashion. J Virol 76:1328–1338

Hansen SG, Strelow LI, Franchi DC, Anders DG, Wong SW (2003) Complete sequence and genomic analysis of rhesus cytomegalovirus. J Virol 77:6620–6636

Harkins L, Volk AL, Samanta M, Mikolaenko I, Britt WJ, Bland KI, Cobbs CS (2002) Specific localisation of human cytomegalovirus nucleic acids and proteins in human colorectal cancer. Lancet 360:1557–1563

Hesselgesser J, Ng HP, Liang M, Zheng W, May K, Bauman JG, Monahan S, Islam I, Wei GP, Ghannam A, Taub DD, Rosser M, Snider RM, Morrissey MM, Perez HD, Horuk R (1998) Identification and characterization of small molecule functional antagonists of the CCR1 chemokine receptor. J Biol Chem 273:15687–15692

Hulshof JW, Casarosa P, Menge WM, Kuusisto LM, van der Goot H, Smit MJ, de Esch IJ, Leurs R (2005) Synthesis and structure activity relationship of the first nonpeptidergic inverse agonists for the human cytomegalovirus encoded chemokine receptor US28. J Med Chem 48:6461–6471

Isegawa Y, Ping Z, Nakano K, Sugimoto N, Yamanishi K (1998) Human herpesvirus 6 open reading frame U12 encodes a functional beta-chemokine receptor. J Virol 72:6104–6112

Jensen KK, Manfra DJ, Grisotto MG, Martin AP, Vassileva G, Kelley K, Schwartz TW, Lira SA (2005) The human herpes virus 8-encoded chemokine receptor is required for angioproliferation in a murine model of Kaposi's sarcoma. J Immunol 174:3686–3694

Kaptein SJ, Beisser PS, Gruijthuijsen YK, Savelkouls KG, van Cleef KW, Beuken E, Grauls GE, Bruggeman CA, Vink C (2003) The rat cytomegalovirus R78 G protein-coupled receptor gene is required for production of infectious virus in the spleen. J Gen Virol 84:2517–2530

Kirshner JR, Staskus K, Haase A, Lagunoff M, Ganem D (1999) Expression of the open reading frame 74 (G-protein-coupled receptor) gene of Kaposi's sarcoma (KS)-associated herpesvirus: implications for KS pathogenesis. J Virol 73:6006–6014

Kledal TN, Rosenkilde MM, Schwartz TW (1998) Selective recognition of the membrane-bound CX3C chemokine, fractalkine, by the human cytomegalovirus-encoded broad-spectrum receptor US28. FEBS Letters 441:209–214

Kuhn DE, Beall CJ, Kolattukudy PE (1995) The cytomegalovirus US28 protein binds multiple CC chemokines with high affinity. Biochem Biophys Res Commun 211:325–330

Landolfo S, Gariglio M, Gribaudo G, Lembo D (2003) The human cytomegalovirus. Pharmacol Ther 98:269–297

Laruelle M, Abi-Dargham A (1999) Dopamine as the wind of the psychotic fire: new evidence from brain imaging studies. J Psychopharmacol 13:358–371

Liang M, Rosser M, Ng HP, May K, Bauman JG, Islam I, Ghannam A, Kretschmer PJ, Pu H, Dunning L, Snider RM, Morrissey MM, Hesselgesser J, Perez HD, Horuk R (2000) Species selectivity of a small molecule antagonist for the CCR1 chemokine receptor. Eur J Pharmacol 389:41–49

Margulies BJ, Browne H, Gibson W (1996) Identification of the human cytomegalovirus G protein-coupled receptor homologue encoded by UL33 in infected cells and enveloped virus particles. Virology 225:111–125

Maussang D, Verzijl D, van Walsum M, Leurs R, Holl J, Pleskoff O, Michel D, van Dongen GA, Smit MJ (2006) Human cytomegalovirus-encoded chemokine receptor US28 promotes tumorigenesis. Proc Natl Acad Sci USA 103:13068–13073

McFadden G (2005) Poxvirus tropism. Nat Rev Microbiol 3:201–213

McGeoch DJ (2001) Molecular evolution of the gamma-Herpesvirinae. Philos Trans R Soc Lond B Biol Sci 356:421–435

McMaster BE, Schall TJ, Penfold M, Wright JJ, Dairaghi DJ (2003a) Arylamines as inhibitors of chemokine binding to US28. World (PTC) Patent WO03020029

McMaster BE, Schall TJ, Penfold M, Wright JJ, Dairaghi DJ (2003b) Bicyclic compounds as inhibitors of chemokine binding to US28. World (PTC) Patent WO03018549

Melnychuk RM, Smith P, Kreklywich CN, Ruchti F, Vomaske J, Hall L, Loh L, Nelson JA, Orloff SL, Streblow DN (2005) Mouse cytomegalovirus M33 is necessary and sufficient in virus-induced vascular smooth muscle cell migration. J Virol 79:10788–10795

Michel D, Milotic I, Wagner M, Vaida B, Holl J, Ansorge R, Mertens T (2005) The human cytomegalovirus UL78 gene is highly conserved among clinical isolates, but is dispensable for replication in fibroblasts and a renal artery organ-culture system. J Gen Virol 86:297–306

Middeldorp JM, Brink AA, van den Brule AJ, Meijer CJ (2003) Pathogenic roles for Epstein-Barr virus (EBV) gene products in EBV-associated proliferative disorders. Crit Rev Oncol Hematol 45:1–36

Milne RS, Mattick C, Nicholson L, Devaraj P, Alcami A, Gompels UA (2000) RANTES binding and down-regulation by a novel human herpesvirus-6 beta chemokine receptor. J Immunol 164:2396–2404

Minisini R, Tulone C, Luske A, Michel D, Mertens T, Gierschik P, Moepps B (2003) Constitutive inositol phosphate formation in cytomegalovirus-infected human fibroblasts is due to expression of the chemokine receptor homologue pUS28. J Virol 77:4489–4501

Mocarski ES, Courcelle CT (2001) Cytomegalovirus and their replication. In: Knipe D, Howley P (eds) Field's virology. Lippincott, Williams and Wilkins, Philadelphia, pp 2629–2673

Mokros T, Rehm A, Droese J, Oppermann M, Lipp M, Hopken UE (2002) Surface expression and endocytosis of the human cytomegalovirus-encoded chemokine receptor US28 is regulated by agonist-independent phosphorylation. J Biol Chem 277:45122–45128

Montaner S, Sodhi A, Pece S, Mesri EA, Gutkind JS (2001) The Kaposi's sarcoma-associated herpesvirus G protein-coupled receptor promotes endothelial cell survival through the activation of Akt/protein kinase B. Cancer Res 61:2641–2648

Montaner S, Sodhi A, Molinolo A, Bugge TH, Sawai ET, He Y, Li Y, Ray PE, Gutkind JS (2003) Endothelial infection with KSHV genes in vivo reveals that vGPCR initiates Kaposi's sarcomagenesis and can promote the tumorigenic potential of viral latent genes. Cancer Cell 3:23–36

Montaner S, Sodhi A, Ramsdell AK, Martin D, Hu J, Sawai ET, Gutkind JS (2006) The Kaposi's sarcoma-associated herpesvirus G protein-coupled receptor as a therapeutic target for the treatment of Kaposi's sarcoma. Cancer Res 66:168–174

Murphy PM (2002) International Union of Pharmacology. XXX. Update on chemokine receptor nomenclature. Pharmacol Rev 54:227–229

Murphy PM, Baggiolini M, Charo IF, Hebert CA, Horuk R, Matsushima K, Miller LH, Oppenheim JJ, Power CA (2000) International union of pharmacology. XXII. Nomenclature for chemokine receptors. Pharmacol Rev 52:145–176

Najarro P, Lee HJ, Fox J, Pease J, Smith GL (2003) Yaba-like disease virus protein 7L is a cell-surface receptor for chemokine CCL1. J Gen Virol 84:3325–3336

Najarro P, Gubser C, Hollinshead M, Fox J, Pease J, Smith GL (2006) Yaba-like disease virus chemokine receptor 7L, a CCR8 orthologue. J Gen Virol 87:809–816

Nakano K, Tadagaki K, Isegawa Y, Aye MM, Zou P, Yamanishi K (2003) Human herpesvirus 7 open reading frame U12 encodes a functional beta-chemokine receptor. J Virol 77:8108–8115

Offermanns S (2003) G-proteins as transducers in transmembrane signalling. Prog Biophys Mol Biol 83:101–130

Oliveira SA, Shenk TE (2001) Murine cytomegalovirus M78 protein, a G protein-coupled receptor homologue, is a constituent of the virion and facilitates accumulation of immediate-early viral mRNA. Proc Natl Acad Sci USA 98:3237–3242

Onuffer JJ, Horuk R (2002) Chemokines, chemokine receptors and small-molecule antagonists: recent developments. Trends Pharmacol Sci 23:459–467

Oster B, Hollsberg P (2002) Viral gene expression patterns in human herpesvirus 6B-infected T cells. J Virol 76:7578–7586

Paulsen SJ, Rosenkilde MM, Eugen-Olsen J, Kledal TN (2005) Epstein–Barr virus-encoded BILF1 is a constitutively active G protein-coupled receptor. J Virol 79:536–546

Penfold ME, Schmidt TL, Dairaghi DJ, Barry PA, Schall TJ (2003) Characterization of the rhesus cytomegalovirus US28 locus. J Virol 77:10404–10413

Pleskoff O, Treboute C, Brelot A, Heveker N, Seman M, Alizon M (1997) Identification of a chemokine receptor encoded by human cytomegalovirus as a cofactor for HIV-1 entry. Science 276:1874–1878

Ray N, Doms RW (2006) HIV-1 coreceptors and their inhibitors. Curr Top Microbiol Immunol 303:97–120

Rosenkilde MM, Kledal TN, Brauner-Osborne H, Schwartz TW (1999) Agonists and inverse agonists for the herpesvirus 8-encoded constitutively active seven-transmembrane oncogene product, ORF-J. Biol Chem 274:956–961

Sahagun-Ruiz A, Sierra-Honigmann AM, Krause P, Murphy PM (2004) Simian Cytomegalovirus Encodes Five Rapidly Evolving Chemokine Receptor Homologues. Virus Genes 28:71–83

Samanta M, Harkins L, Klemm K, Britt WJ, Cobbs CS (2003) High prevalence of human cytomegalovirus in prostatic intraepithelial neoplasia and prostatic carcinoma. J Urol 170:998–1002

Schall TJ, McMaster BE, Dairaghi DJ (2002a) Modulators of USWorld (PTC) Patent WO0217900

Schall TJ, McMaster BE, Dairaghi DJ (2002b) Reagents and methods for the diagnosis of CMV dissemination. World (PTC) Patent WO0217969

Smit MJ, Verzijl D, Casarosa P, Navis M, Timmerman H, Leurs R (2002) Kaposi's sarcoma-associated herpesvirus-encoded G protein-coupled receptor ORF74 constitutively activates p44/p42 MAPK and Akt via G(i) and phospholipase C-dependent signaling pathways. J Virol 76:1744–1752

Soderberg-Naucler C (2006) Does cytomegalovirus play a causative role in the development of various inflammatory diseases and cancer? J Intern Med 259:219–246

Sodhi A, Montaner S, Patel V, Zohar M, Bais C, Mesri EA, Gutkind JS (2000) The Kaposi's sarcoma-associated herpes virus G protein-coupled receptor up-regulates vascular endothelial growth factor expression and secretion through mitogen-activated protein kinase and p38 pathways acting on hypoxia-inducible factor 1alpha. Cancer Res 60:4873–4880

Sodhi A, Montaner S, Gutkind JS (2004a) Viral hijacking of G-protein-coupled-receptor signalling networks. Nat Rev Mol Cell Biol 5:998–1012

Sodhi A, Montaner S, Patel V, Gomez-Roman JJ, Li Y, Sausville EA, Sawai ET, Gutkind JS (2004b) Akt plays a central role in sarcomagenesis induced by Kaposi's sarcoma herpesvirus-encoded G protein-coupled receptor. Proc Natl Acad Sci USA 101:4821–4826

Stassen FR, Vega-Cordova X, Vliegen I, Bruggeman CA (2006) Immune activation following cytomegalovirus infection: more important than direct viral effects in cardiovascular disease? J Clin Virol 35:349–353

Streblow DN, Nelson JA (2003) Models of HCMV latency and reactivation. Trends Microbiol 11:293–295

Streblow DN, Soderberg-Naucler C, Vieira J, Smith P, Wakabayashi E, Ruchti F, Mattison K, Altschuler Y, Nelson JA (1999) The human cytomegalovirus chemokine receptor US28 mediates vascular smooth muscle cell migration. Cell 99:511–520

Streblow DN, Kreklywich CN, Smith P, Soule JL, Meyer C, Yin M, Beisser P, Vink C, Nelson JA, Orloff SL (2005) Rat cytomegalovirus-accelerated transplant vascular sclerosis is reduced with mutation of the chemokine-receptor R33. Am J Transplant 5:436–442

Sun R, Lin SF, Staskus K, Gradoville L, Grogan E, Haase A, Miller G (1999) Kinetics of Kaposi's sarcoma-associated herpesvirus gene expression. J Virol 73:2232–2242

Tadagaki K, Nakano K, Yamanishi K (2005) Human herpesvirus 7 open reading frames U12 and U51 encode functional beta-chemokine receptors. J Virol 79:7068–7076

Thompson MP, Kurzrock R (2004) Epstein-Barr virus and cancer. Clin Cancer Res 10:803–821

Thorley-Lawson DA (2005) EBV the prototypical human tumor virus–just how bad is it? J Allergy Clin Immunol 116:251–261

Thorley-Lawson DA, Gross A (2004) Persistence of the Epstein-Barr virus and the origins of associated lymphomas. N Engl J Med 350:1328–1337

Verzijl D, Fitzsimons CP, Van Dijk M, Stewart JP, Timmerman H, Smit MJ, Leurs R (2004) Differential activation of murine herpesvirus 68- and Kaposi's sarcoma-associated herpesvirus-encoded ORF74 G protein-coupled receptors by human and murine chemokines. J Virol 78:3343–3351

Vieira J, Schall TJ, Corey L, Geballe AP (1998) Functional analysis of the human cytomegalovirus US28 gene by insertion mutagenesis with the green fluorescent protein gene. J Virol 72:8158–8165

Vischer HF, Leurs R, Smit MJ (2006a) HCMV-encoded G-protein-coupled receptors as constitutively active modulators of cellular signaling networks. Trends Pharmacol Sci 27:56–63

Vischer HF, Vink C, Smit MJ (2006b) A viral conspiracy: hijacking the chemokine system through virally encoded pirated chemokine receptors. Curr Top Microbiol Immunol 303:121–154

Waldhoer M, Kledal TN, Farrell H, Schwartz TW (2002) Murine cytomegalovirus (CMV) M33 and human CMV US28 receptors exhibit similar constitutive signaling activities. J Virol 76:8161–8168

Wang F, Rivailler P, Rao P, Cho Y (2001) Simian homologues of Epstein–Barr virus. Philos Trans R Soc Lond B Biol Sci 356:489–497

Ernst Schering Foundation Symposium Proceedings, Vol. 2, pp. 211–228
DOI 10.1007/2789_2006_010

Published Online: 16 May 2007

Modulation of GPCR Conformations by Ligands, G-Proteins, and Arrestins

E.R. Prossnitz(✉), L.A. Sklar

Department of Cell Biology and Physiology, Cancer Research and Treatment Center, University of New Mexico Health Sciences Center, 87131 Albuquerque NM, USA
email: *eprossnitz@salud.unm.edu*

Abstract. G-protein-coupled receptors (GPCRs) have traditionally been thought to adopt two conformations: the inactive unliganded conformation and the active ligand-bound conformation. Interactions with G-proteins in cells and membranes are known to modulate the affinity of the receptor for ligand and therefore the conformation of the receptor. Such observations led to the proposal of the ternary complex model. However, subsequent studies of constitutively active GPCRs led to the development of an extended version of this model to account for active conformations of the receptor in the absence of agonist. A significant difficulty with many of the studies, upon which this latter model was based, is the lack of knowledge of receptor and G-protein concentrations due to the two-dimensional nature of the membranes used to perform the measurements. Over the past decade, we have studied the interaction of GPCRs, G-proteins, arrestins, and ligands in solubilized systems, where the concentration of each component can be defined. Here we summarize results of these studies as they pertain to the regulation of GPCR conformations and affinities for interacting species.

1 Introduction

G-protein-coupled receptors (GPCRs) represent the largest single class of targets for therapeutic drugs (Wise et al. 2002), as they are involved in the regulation of physiological processes in every organ of the body (Hill 2006). As such they respond to a plethora of ligands that include ions, amino acids, peptides (including (glyco)proteins and proteases), monoamines, lipids, nucleotides, and light and small organic molecules (olfactants). Upon binding of an agonist, the receptor stimulates one or more heterotrimeric G-proteins, which leads to activation of downstream effectors (Maudsley et al. 2005). This is followed by receptor phosphorylation, mediated primarily by G-protein-coupled receptor kinases (GRKs), and the subsequent binding of arrestins. At each step in these processes, the possibility exists for a large number of interactions from which high levels of signaling specificity may be derived. For example, one GPCR can be activated by many ligands (both natural and synthetic), leading to differential cellular responses. Furthermore, the presence of a diverse array of possible G-protein heterotrimers as well as multiple GRK and arrestin subtypes leads to a dizzying complexity in the potential number of protein complexes (Vauquelin and Van Liefde 2005). Although it was originally postulated that all GPCR signaling proceeded through G-proteins and that receptor phosphorylation and arrestin binding provided solely for desensitization, it is now clear that arrestins can also act as scaffolding molecules, leading to the initiation and regulation of signaling and receptor trafficking (Luttrell 2005; Prossnitz 2004).

To further the understanding of GPCR-mediated processes and interactions, we have utilized the *N*-formyl peptide receptor (FPR), which couples to a pertussis toxin-sensitive Gi protein and is expressed predominantly on neutrophils (Le et al. 2002; Prossnitz and Ye 1997). This receptor recognizes bacterially generated *N*-formyl peptides that act as potent chemoattractants for human phagocytes. The FPR is one of the better-characterized receptors in the chemoattractant/chemokine subclass of GPCRs, modulating several cell functions including chemotaxis, superoxide formation, degranulation, and transcription. Extensive biophysical characterization of the FPR has been carried out since the early 1980s, in large part due to the ease of synthesis and modification of its ligand (Schmitt et al. 1983; Sklar et al. 1981). In addition to iodinated ligands (Schmitt et al. 1983), formylated peptides have also been modified to incorporate fluorescent groups such as fluorescein and rhodamine (Niedel et al. 1980). These fluorescent groups can be placed at almost any position within the ligand, generating entire families of ligands of varying composition and affinity (Vilven et al. 1998). Cell-based assays were employed in conjunction with the fluorescent ligands to probe both the physical and biological properties of the receptor (Sklar et al. 1984). In the 1990s, such techniques were extended to FPR mutants to examine the processing of the receptor. In the late 1990s, we began to develop procedures to solubilize the FPR (Sklar et al. 2000), allowing for more detailed studies of receptor function through the reconstitution of receptor-G-protein and receptor-arrestin complexes (Bennett et al. 2001a,b; Key et al. 2001, 2003, 2005). In order to probe mechanisms of partial agonists, we subsequently developed bead-based assays permitting flow cytometric analysis of receptor assemblies (Simons et al. 2003, 2004). In the following chapter, we summarize the novel mechanisms and interactions that have been characterized using these systems as well as the conclusions drawn from these experiments.

2 Molecular Assays of Solubilized GPCR Complexes

Many of the models of GPCR function are based primarily on physiological and cellular data as much less information is available on the biophysical interactions of GPCRs. In order to study the interactions of GPCRs with their primary interacting partners, namely ligands, G-proteins, and arrestins, we undertook experiments using solubilized receptors and purified proteins (Sklar et al. 2002). In studies beginning in the 1980s, we described real-time fluorescence and flow cytometric assays of ligand–receptor interactions that were primarily directed toward viable intact cells or detergent-permeabilized whole cells (Fay et al. 1991; Sklar et al. 1981, 1985, 1987, 1989). Such studies characterized first wild-type and then mutant receptors leading to a model of receptor–G-protein coupling for the FPR (Bennett et al. 2001b; Key et al. 2001, 2003; Prossnitz et al. 1999). In subsequent studies, we were able to determine the efficiency of FPR solubilization using fluorescence methods with epitope-tagged receptors and affinity beads (Sklar et al. 2000). These studies provided the first evidence that the solubilized FPR was able to reconstitute with G-proteins in cell-free and membrane-free systems (Bennett et al. 2001b).

Receptor–G-protein interactions have now been characterized in intact and permeabilized cells, membranes, reconstituted phospholipid vesicles, and in detergent using sucrose gradients. Based on observations with intact and permeabilized cells, we extended our approach to detergent-solubilized receptors where we could independently monitor ligand binding as well as receptor–G-protein binding events. The approach takes advantage of the fact that the ligand dissociation rate of the ligand-receptor (LR) complex of the FPR (in the absence of G-protein or in the presence of GTPγS) is many times faster than the ligand dissociation from the G protein-coupled complex (LRG) (Sklar et al. 1987). This is consistent with the known properties of GPCRs, where the interaction of a G-protein results in a higher affinity of the receptor for ligand. When GTP is not present, the G-protein remains bound to the FPR. However, upon the addition of GTP, or its nonhydrolyzable analog GTPγS, dissociation of the G protein from the receptor is induced, resulting in a decrease in the affinity for the ligand, which is observed as an increase in its dissociation rate. The change in the ligand dissociation rate represents a switch from the slowly dissociating receptor–

Fig. 1 a–c. a The assay of detergent (DOM, dodecyl-maltoside)-solubilized receptors using fluorescent ligands (*N*-fpep*) and anti-FITC antibodies. Interactions with G-proteins and arrestins modulate the dissociation rate of ligand from receptor. Free ligand is rapidly bound and quenched by the anti-FITC antibody. **b** Representative data showing the difference in ligand dissociation rates between free receptor and receptor bound to G-proteins following the addition of anti-FITC antibodies. Addition of GTPγS where indicated demonstrates the conversion of receptor from high ligand affinity to low ligand affinity. **c** Two configurations of bead-based assemblies. In the *upper model*, heterotrimeric G-proteins are attached to beads via an epitope tag. Receptor-GFP chimeric proteins are capable of associating with the G-protein-coated bead only in the presence of agonists. In the *lower model*, ligand is covalently attached to the bead, which can then directly bind receptor-GFP proteins

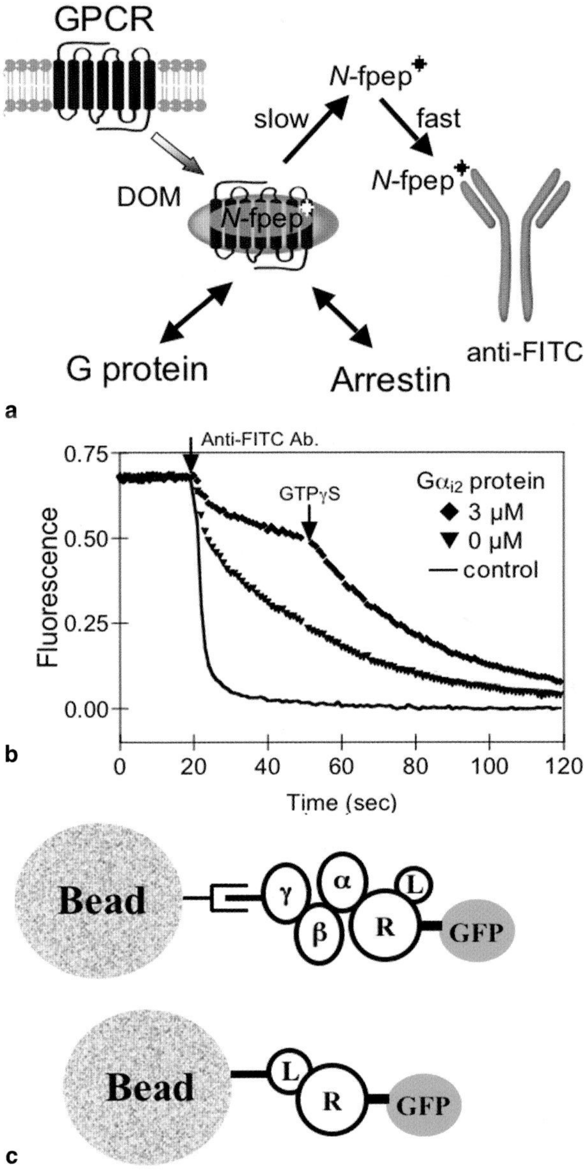
GPCR
N-fpep
slow
fast
DOM
N-fpep
N-fpep
G protein
Arrestin
anti-FITC
a
Anti-FITC Ab.
GTPγS
Gα$_{i2}$ protein
3 μM
0 μM
control
Fluorescence
0.75
0.50
0.25
0.00
b
0
20
40
60
80
100
120
Time (sec)
Bead
γ
α
L
β
R
GFP
Bead
L
R
GFP
c

G-protein complex to the rapidly dissociating uncomplexed receptor as the activated G-protein functionally dissociates from the receptor.

The assay is based upon a fluorescein-conjugated ligand (*N*-formyl-Met-Leu-Phe-Lys-fluorescein 5-isothiocyanate, fMLFK-FITC) used in conjunction with an antifluorescein antibody that rapidly quenches the fluorescence of free fluorescein (FITC), as shown in Fig. 1a and b (Bennett et al. 2001b). The antibody binds to the fluorescein on the free ligand (i.e., not bound to receptor), as FPR–ligand complexes sterically inhibit the antibody from accessing the fluorescein on this particular short peptide. In this manner, we can determine the quantity of FPR-bound ligand immediately upon the addition of the antifluorescein antibody. Furthermore, we can follow the ligand dissociation kinetics from the receptor as the excess antibody quenches the ligand being released from the receptor. The dissociation data can be fit to one or two rates, the slow rate representing ligand dissociating from receptor–G-protein complex and the fast rate representing ligand dissociating from uncomplexed receptor.

In contrast to the homogenous solution-based methods described above, we have also developed bead-based methods where one of the components of the system is attached or bound to a bead and the binding of other (fluorescently tagged) components is determined by flow cytometry. Two examples of such systems are shown in Fig. 1c. The upper scheme is used to analyze receptor–G-protein interactions, whereas the second is used to measure receptor–ligand interactions (Simons et al. 2003). In the upper scheme, the βγ subunit of the G-protein is immobilized on the bead via either a hexahistidine or FLAG tag on the carboxy terminus of the γ subunit. In the former case, nickel chelate beads are used; in the second, streptavidin beads coated with biotinylated anti-FLAG antibodies are used. To these complexes, a purified G-protein alpha subunit can then be bound to generate beads containing the heterotrimer. The fluorescent receptor then consists of either a solubilized receptor–GFP fusion protein that binds to G-protein-coated beads only in the presence of agonist (Fig. 1c, upper scheme). In an alternate format, native receptor is incubated in the presence of a fluorescent ligand (not shown). The former scheme permits direct monitoring of the bound receptor, whereas the latter allows indirect determination of bound receptor via its associated fluorescent ligand. Interactions between receptor (receptor–GFP fusions) and ligands can also be directly monitored in the lower scheme of Fig. 1c using beads to which a ligand has been covalently attached. In such a configuration, competition by ligand results in the loss of bead-based fluorescence.

3 Reconstitution with Soluble G-Proteins

In our initial studies, we focused our investigation on the interaction of the detergent-solubilized FPR with G-proteins in solution. Using this reconstitution system, we demonstrated the ability of the receptor to couple with endogenous G-proteins, as well as with exogenous G-proteins containing specific Gα subunits (Bennett et al. 2001b). We were also able to measure the affinities of the complexes and found that the FPR binds to a G-protein heterotrimer containing the Gai3 subunit with somewhat higher affinity than to heterotrimers containing Gαi2 or Gαi1 proteins. The individual α subunits and the βγ complex alone were unable to induce a change in the dissociation rate of ligand from the receptor indicating a necessity for the presence of the intact G-protein heterotrimer. We were also able to inhibit the G-protein–receptor interaction with peptides derived from the Gαi subunits as well as with anti-Gαi antibodies.

To further investigate the interactions of GPCRs with G-proteins, we solubilized the FPR from cells, which had been stimulated with a saturating concentration of a low-affinity ligand (f-Met-Leu-Phe) for 10 min prior to harvesting of the cells (Key et al. 2001). The goal of this treatment was to obtain endogenously phosphorylated FPR for solubilization and subsequent reconstitution. Incubation of fluorescent ligand with the solubilized FPR prepared from unstimulated cells leads to the formation of slowly dissociating, nucleotide-sensitive complexes as described above. However, in the case of the FPR obtained from fMLF-stimulated cells (i.e., phosphorylated FPR), there was also evidence for the time-dependent formation of a slowly dissociating complex. This complex, with high ligand binding affinity, was GTPγS-insensitive (both in the short term following antibody addition and upon incubation with GTPγS during the reconstitution phase), unlike the well-characterized receptor–G-protein complex. Thus, although there was evidence of time-dependent complex formation in the case of phosphorylated receptor, its precise makeup was not immediately clear. Given the GTPγS insensitivity of the complex, it was likely that the assembly did not involve GTP-binding proteins.

In the context of these results, we then examined the effects of immunodepletion of both endogenous G-proteins and arrestins on the ligand dissociation kinetics for both the nonphosphorylated and phosphorylated solubilized receptor preparations (Key et al. 2001). As we had previously demonstrated, immunodepletion of Gi proteins from the nonphosphorylated receptor preparation completely inhibited the formation of the high-affinity, nucleotide-sensitive LRG complex. However, G-protein depletion from the phosphorylated receptor preparation had a negligible effect on the time-dependent agonist affinity changes. Thus, G-protein depletion modulated receptor ligand affinity in a similar manner to GTPγS pretreatment for the unphosphorylated FPR. Arrestin depletion, on the other hand, displayed little to no effect on the ligand dissociation kinetics of the nonphosphorylated FPR complex, but completely prevented the formation of a high-affinity ligand-binding complex formed by the phosphorylated FPR. Furthermore, it appeared that arrestin-depletion mildly enhances the nucleotide sensitivity of phosphorylated FPR complexes, suggesting a low level of receptor–G-protein complex formation in the absence of arrestin. In addition, the simultaneous depletion of both G-proteins and arrestins prevented both the characteristic receptor–G-protein complex as well as the high-affinity, nucleotide-insensitive complex formed by the nonphosphorylated and phosphorylated FPR preparations, respectively. Therefore, removal of both endogenous proteins appeared to be sufficient to prevent all of the time-dependent ligand affinity changes regardless of the phosphorylation state of the receptor, indicating the lack of additional interacting proteins. Finally, in the case of the phosphorylated receptor, reconstitution with exogenous G-protein, even at three times the EC_{50} concentration for the nonphosphorylated FPR, resulted in neither high ligand affinity complexes nor nucleotide-sensitive ligand dissociation, consistent with the absence of complex formation.

In order to examine whether the phosphorylated form of the FPR was capable of interacting with G proteins when not phosphorylated, we examined the ability of alkaline phosphatase to convert the agonist dissociation kinetics and associated protein coupling characteristics of the activated, phosphorylated receptor to that of the unstimulated, nonphosphorylated receptor (Key et al. 2001). For this, solubilized membranes were incubated with alkaline phosphatase prior to ligand addition and reconstitution. The results demonstrated that the phosphatase-pretreated, stimulated receptor behaves in a similar manner to the native untreated, nonphosphorylated receptor. Hence, incubation with exogenous G-proteins but not arrestin resulted in the formation of a high-affinity, nucleotide-sensitive complex characteristic of receptor–G-protein complexes. These results were the first to demonstrate in a solubilized system that the phosphorylated form of a GPCR exhibited reduced affinity for G-proteins, while simultaneously increasing the affinity of the receptor for arrestin.

Since "complete" phosphorylation of the FPR, as defined by the extent that occurs upon receptor stimulation in cells, inhibited the interaction of the receptor with G-proteins, we sought to examine the effects of phosphorylation in greater detail (Bennett et al. 2001a). We had previously characterized the cellular functions of several FPR phosphorylation-deficient mutants, replacing potential phosphorylation sites (serine or threonine residues) with either alanine or glycine residues (Maestes et al. 1999). These included mutant A (S328A, T329A, T331A, S332G) and mutant B (T334G, T336G, S338G, T339A), which represent mutations of two clusters of Ser and Thr residues (the A site and the B site, respectively). Mutants C (S328A, T329A) and D (T331A, S332G) each restore two potential phosphorylation sites altered in mutant A. Previous characterization of these mutants had demonstrated that they induce calcium mobilization and internalize to the same extent as the wild-type receptor, indicating that their signal initiating properties were unaffected by the mutations. However, they did differ in their overall levels of phosphorylation and their ability to desensitize over time (Maestes et al. 1999). Reconstitution studies revealed that in their unphosphorylated state (from unstimulated cells), all four receptors were capable of producing a high-ligand-affinity complex in the presence of exogenous G-proteins, consistent with their ability to initiate signaling in cells (Bennett et al. 2001a).

To examine the effect of phosphorylation on the mutant FPR–G-protein interactions, we used membrane preparations from cells that had been treated with agonist prior to preparation, as described above. Receptor preparations were immunodepleted of endogenous arrestins, since arrestins bind to phosphorylated receptors and would compete with G-protein coupling. Whereas the phosphorylated WT receptor was unable to couple to G-protein as expected, both the A and B receptors displayed high ligand affinity indicative of G-protein coupling, with the ligand dissociation rate being sensitive to addition of GTPγS (Bennett et al. 2001a). On the contrary, the phosphorylated C and D receptors were indistinguishable from phosphorylated wild-type receptors. These results demonstrated that there may be a threshold of receptor phosphorylation above which G-protein binding is directly inhibited. Whether and to what extent this occurs for other GPCRs remains to be determined.

4 Reconstitution with Arrestins

As alluded to above, stimulation of cells expressing the FPR resulted in the generation of a high-affinity ligand-binding complex that was insensitive to the addition of GTPγS. Furthermore, formation of this complex could be prevented by immunodepletion of arrestin. To examine these interactions in greater detail, we utilized reconstitution studies with purified arrestins much in the same was as we used purified G-proteins in the previous section. Our results demonstrated that both arrestin-2 and arrestin-3 were capable of interacting with the phosphorylated FPR to create a high-ligand-affinity, GTPγS-insensitive complex (Key et al. 2001). Neither protein had an effect on the affinity of unphosphorylated receptor. To test whether the phosphorylation state of the FPR was solely responsible for the formation of the receptor–arrestin complex, we treated solubilized receptors with alkaline phosphatase as above. Whereas such treatment restored the ability of phosphorylated receptors to interact with G-proteins, with respect to arrestin reconstitution, it completely prevented any complex formation. Finally, because our system could not directly preclude the existence of a complex in the absence of a change in ligand affinity, we investigated whether preincubation of solubilized receptor with either exogenous G-proteins or arrestins would prevent subsequent agonist affinity shifts associated with the other protein. In the case of the phosphorylated receptor, high concentrations of G-proteins did not inhibit agonist affinity changes resulting from the addition of arrestin. Similarly with nonphosphorylated FPR, high concentrations of arrestin did not prevent agonist-dependent affinity changes resulting from G-protein reconstitution, indicating that an absence of the ligand affinity change was consistent with the lack of protein binding to the receptor.

To extend these results with respect to the level of receptor phosphorylation, we utilized the same series of partially phosphorylated FPR mutants described above. Like the wild-type phosphorylated receptors that displayed a high affinity for ligand upon incubation with purified arrestin, the mutant C and D receptors, but not the mutant A and B receptors, were also able to induce such a state (Bennett et al. 2001a). To further examine whether the lack of the high-ligand-affinity state in the mutant A and B receptors was due to an inability to bind arrestin or due to the inability of the receptor-bound arrestin to induce an affinity change, competition experiments were conducted. Because the phosphorylated mutant A and B receptors were able to bind G-protein, we attempted to block the binding of G-protein with excess arrestin. The lack of any competition indicated that these two mutants were not able to bind arrestin. In addition to reconstitution experiments, we examined the ability of arrestin to colocalize in vivo with the phosphorylation-deficient mutants using confocal fluorescence microscopy. Cells expressing either WT or mutant receptors were stimulated with a fluorescent agonist to track the FPR as it was processed and internalized. Arrestins were detected using an anti-arrestin antibody. The results demonstrated that the mutant C and D receptors, like the wild-type receptor, associated with arrestin following internalization. However, although the mutant A and B receptors also internalize following stimulation, there was no colocalization with arrestin. These results are entirely consistent with the reconstitution studies and indicate that in addition to a phosphorylation threshold for the inhibition of G-protein binding, a similar (inverse) threshold exists for the binding of arrestins.

Although arrestins bind selectively to phosphorylated ligand-bound GPCRs, truncation of the carboxy-terminal tail of arrestins has been shown to produce a preactivated protein. It is hypothesized that the tail of arrestin stabilizes the basal inactive state by intramolecular interactions with the polar core domain. Truncation is believed to reduce the activation energy of arrestin by the removal of stabilizing interactions that maintain the molecule in an inactive conformation. Arrestin truncation has been demonstrated to lead to partial phosphorylation-independent as well as activation (i.e. ligand-)independent interactions with a number of GPCRs, including receptors from which the carboxyl terminus, containing critical phosphorylation sites, has been removed.

Reconstitution of phosphorylated FPR with truncated arrestin-2 resulted in a higher fraction of slowly dissociating ligand as compared to reconstitution with wild-type arrestin (Key et al. 2003). Titration studies of truncated arrestin-2 with the phosphorylated FPR demonstrate an EC_{50} of 220 nM compared to wild-type arrestin-2, which displayed an EC_{50} of 600 nM. Interestingly, reconstitution in the absence of ligand (with ligand added 5 min prior to assay to load the receptor) failed to result in the production of a complex, suggesting that the truncated arrestin did not bind to the nonliganded state of the receptor. In addition, incubation of the nonphosphorylated FPR with truncated arrestin-2 had no discernible effects on ligand affinity, in contrast to results with the phosphorylated FPR. However, competition studies of G-protein binding to nonphosphorylated receptor demonstrated that truncated arrestin-2 is capable of binding to the FPR with low affinity.

Since truncated arrestin-2 was capable of interacting with both unphosphorylated and phosphorylated receptor, we asked what the effects of partial receptor phosphorylation might be (Key et al. 2003). Whereas the partially phosphorylated mutant A receptor did not bind to wild-type arrestin, it did exhibit high ligand affinity upon reconstitution with truncated arrestin-2. The affinity of the interaction was, however, approximately tenfold lower than that seen with the wild-type arrestin. This result suggests that certain receptor phosphorylation sites are required to activate arrestin for receptor binding and are also involved in determining the affinity of the interaction. Because phosphorylation of the mutant A receptor does not prohibit G-protein coupling, we could also demonstrate the successful competition of G-protein binding with the truncated arrestin-2, as suggested by the loss of nucleotide-sensitivity of the high-ligand-affinity complex. In contrast to the mutant A receptor, incubation of the phosphorylated mutant B receptor with truncated arrestin-2 had no discernible effects on agonist affinity. To differentiate between binding without an accompanying agonist affinity shift and an absolute lack of binding, we carried out competition studies with G-proteins. Reconstitution of the phosphorylated mutant B receptor with exogenous G-proteins resulted in the formation of a high-agonist-affinity, nucleotide-sensitive complex, as described above. However, the addition of truncated arrestin-2 resulted in the conversion of the receptor from a high-affinity to low-agonist-affinity state, as might be predicted from the lack of a ligand-affinity change in the direct binding assay. This competition occurred with an EC_{50} of 340 nM, a similar affinity to the interaction of wild-type arrestin-2 with the phosphorylated, wild-type FPR. Thus, although phosphorylation in the A site is not sufficient for arrestin-dependent agonist affinity changes, it can provide for a high-affinity interaction of arrestin with the FPR. These results were confirmed in vivo using confocal fluorescence microscopy to demonstrate that both the mutant A and mutant B receptors colocalized with truncated arrestin-2 into punctate structures, in contrast to the results with wild-type arrestin-2. These results confirmed that the in vitro reconstitution results were supported by studies of the receptors and arrestin in a native cellular environment.

5 Assessing Partial Agonism with Solubilized GPCRs

Having examined the modulation of affinity of receptor complexes for ligand (i.e., receptor conformation) as a result of receptor phosphorylation and interactions with G-proteins and arrestins, we turned our attention to the issue of partial agonism displayed by many ligands (Simons et al. 2003, 2004). As model systems, we used the FPR and the well-studied beta2-adrenergic receptor (β2AR). For the FPR we examined the affinity of a series of a fluorescent ligands for the soluble receptor as well as the ability of the soluble receptor to assemble with the heterotrimeric G-protein complex on the bead as shown in Fig. 1c. For the β2AR-GFP fusion protein (β2AR-GFP), we were able to compare the affinity of soluble ternary complex assemblies with the affinity of a series of full and partial agonists in competition studies with beads displaying dihydroalprenolol (DHA beads) (Simons et al. 2004).

As shown in Fig. 2a, the binding of β2AR-GFP to DHA beads was sensitive to the presence of agonist (ISO, EPI, NE, SAL, DOB) or antagonist (ALP) and measured ligand affinity. The G-protein–bead assembly, shown in Fig. 2b, was sensitive to the presence of agonists and resolved partial (SAL, DOB) and full agonists (ISO, EPI, NE). The behavior of the complexes on particles was essentially identical to the membrane bound form of the receptor, indicating that the physiological structure of β2AR is retained after solubilization. By modifying one set of beads to fluoresce red or using a different size to resolve the two assays, we could discriminate agonist and antagonist binding in a single step. While a single-step primary screen would resolve agonists and antagonists, a dose-response secondary screen would resolve full and partial agonists. Molecules that specifically block interactions between β2AR and G-protein would not be active in the ligand–bead assembly but would compete with agonist in the G-protein–bead assembly. The ternary complex involving GPCR, G-protein and agonist is formally described by a four-sided model (Fig. 2c). The combination of equilibrium ligand binding and G-protein assembly measurements, as shown for β2AR, is precisely the type of measurement required to resolve ternary complex details. They provide information for the affinity of L binding to R and L binding to RG. One additional required piece of information, G binding to LR, is obtained by using the beads as a sensor, allowing soluble G-protein to compete for liganded GPCR with G-proteins on a bead. With this information, we have now analyzed the four equilibrium constants in the ternary complex for each receptor-ligand complex.

Historically, the ternary complex analysis is particularly challenging because data from GPCRs reflects a combination of 2D and 3D (in the plane and out of the plane of the membrane) affinities and concentrations. By performing the analysis on soluble GPCRs, the calculation is entirely based on 3D parameters. We have fit the data for full agonists to the ternary complex model using *Mathematica* to calculate the series of nonlinear equations. Figure 2d shows the simultaneous fit of all the data and Fig. 2e shows the resulting computed values for the model. The results indicate that partial agonists (SAL, DOB) cause a reduced ternary complex affinity (Kga). This reduction in the affinity of the receptor for G-protein when bound to partial agonists is consistent with the idea that

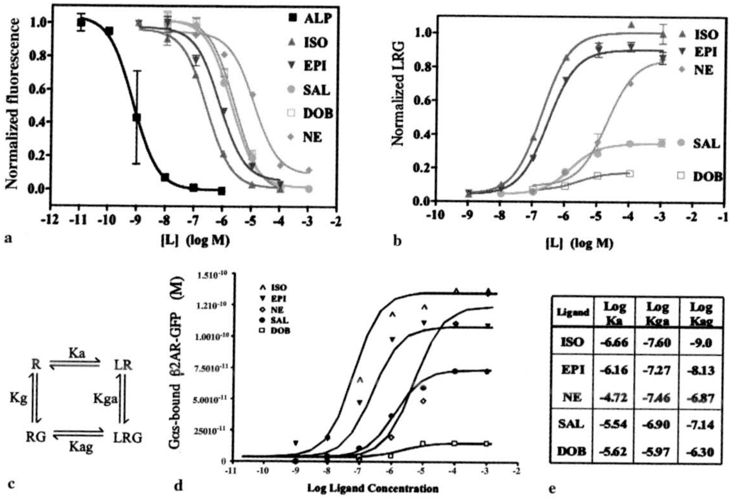

Ligand	Log Ka	Log Kga	Log Kag
ISO	-6.66	-7.60	-9.0
EPI	-6.16	-7.27	-8.13
NE	-4.72	-7.46	-6.87
SAL	-5.54	-6.90	-7.14
DOB	-5.62	-5.97	-6.30

Fig. 2 a–e. a Representative data showing the inhibition of β2-adrenergic receptor-GFP to DHA-coated beads by various ligands. *ALP*, alprenolol; *ISO*, isoproterenol; *EPI*, epinephrine; *SAL*, salbutamol; *DOB*, dobutamine; *NE*, norepinephrine). **b** Representative data showing the binding of β2-adrenergic receptor-GFP to Gαs-coated beads in the presence of the indicated agonists and partial agonists. **c** Representation of the simple ternary complex model of ligand (*L*), receptor (*R*) and G-protein (*G*) with associated affinity constants (*K*). **d** Mathematical fitting of the data in *panels a* (providing Ka) and *b* (providing Kga) to the simple ternary complex. Kg is estimated based upon the lack of RG complex formation using the highest possible G-protein concentrations in the absence of added ligand. Data are expressed as the concentration of β2-adrenergic receptor-GFP bound to the bead-associated Gαs. **e** Calculated parameters of Ka, Kga, and Kga for the various ligands tested

partial agonists may elicit unique receptor conformations with respect to G-protein binding.

6 Conclusions

GPCRs appear to be capable of existing in a large number of distinct conformations, some of these induced by their ligands and others induced by posttranslational modifications (e.g. phosphorylation) or interactions with their intracellular protein-binding partners. We have examined the effects of receptor phosphorylation on binding of G-proteins and arrestins and the resulting effects on GPCR conformation as revealed by changes in ligand affinity. Furthermore, we have examined the effects of partial agonists and full agonists on G-protein binding with respect to the ternary complex model. Overall our results demonstrate that GPCRs do in fact exist in a myriad of conformations, each of which may represent a novel target for drug discovery and therapeutic intervention.

Acknowledgements. Work in the authors' laboratories was supported by grants AI36357 and AI43932 (ERP) and EB00264 (LAS) from the National Institutes of Health, U.S. Public Health Service. We also appreciate the contribution of Dr. Anna Waller to computations efforts.

References

Bennett TA, Foutz TD, Gurevich VV, Sklar LA, Prossnitz ER (2001a) Partial phosphorylation of the N-formyl peptide receptor inhibits G protein association independent of arrestin binding. J Biol Chem 276:49195–49203

Bennett TA, Key TA, Gurevich VV, Neubig R, Prossnitz ER, Sklar LA (2001b) Real-time analysis of G protein-coupled receptor reconstitution in a solubilized system. J Biol Chem 276:22453–22460

Fay SP, Posner RG, Swann WN, Sklar LA (1991) Real-time analysis of the assembly of ligand, receptor, and G protein by quantitative fluorescence flow cytometry. Biochemistry 30:5066–5075

Hill SJ (2006) G-protein-coupled receptors: past, present and future. Br J Pharmacol 147(Suppl 1):S27–37

Key TA, Bennett TA, Foutz TD, Gurevich VV, Sklar LA, Prossnitz ER (2001) Regulation of formyl peptide receptor agonist affinity by reconstitution with arrestins and heterotrimeric g proteins. J Biol Chem 276:49204–49212

Key TA, Foutz TD, Gurevich VV, Sklar LA, Prossnitz ER (2003) N-formyl peptide receptor phosphorylation domains differentially regulate arrestin and agonist affinity. J Biol Chem 278:4041–4047

Key TA, Vines CM, Wagener BM, Gurevich VV, Sklar LA, Prossnitz ER (2005) Inhibition of chemoattractant N-formyl peptide receptor trafficking by active arrestins. Traffic 6:87–99

Le Y, Murphy PM, Wang JM (2002) Formyl-peptide receptors revisited. Trends Immunol 23:541–548

Luttrell LM (2005) Composition and function of G protein-coupled receptor signalsomes controlling mitogen-activated protein kinase activity. J Mol Neurosci 26:253–264

Maestes DC, Potter RM, Prossnitz ER (1999) Differential phosphorylation paradigms dictate desensitization and internalization of the N-formyl peptide receptor. J Biol Chem 274:29791–29795

Maudsley S, Martin B, Luttrell LM (2005) The origins of diversity and specificity in G protein-coupled receptor signaling. J Pharmacol Exp Ther 314:485–494

Niedel J, Kahane I, Lachman L, Cuatrecasas P (1980) A subpopulation of cultured human promyelocytic leukemia cells (HL-60) displays the formyl peptide chemotactic receptor. Proc Natl Acad Sci USA 77:1000–1004

Prossnitz ER (2004) Novel roles for arrestins in the post-endocytic trafficking of G protein-coupled receptors. Life Sci 75:893–899

Prossnitz ER, Ye RD (1997) The N-formyl peptide receptor: a model for the study of chemoattractant receptor structure and function. Pharmacol Ther 74:73–102

Prossnitz ER, Gilbert TL, Chiang S, Campbell JJ, Qin S, Newman W, Sklar LA, Ye RD (1999) Multiple activation steps of the N-formyl peptide receptor. Biochemistry 38:2240–2247

Schmitt M, Painter RG, Jesaitis AJ, Preissner K, Sklar LA, Cochrane CG (1983) Photoaffinity labeling of the N-formyl peptide receptor binding site of intact human polymorphonuclear leukocytes. A label suitable for following the fate of the receptor-ligand complex. J Biol Chem 258:649–654

Simons PC, Shi M, Foutz T, Cimino DF, Lewis J, Buranda T, Lim WK, Neubig RR, McIntire WE, Garrison J, Prossnitz E, Sklar LA (2003) Ligand-receptor-G-protein molecular assemblies on beads for mechanistic studies and screening by flow cytometry. Mol Pharmacol 64:1227–1238

Simons PC, Biggs SM, Waller A, Foutz T, Cimino DF, Guo Q, Neubig RR, Tang WJ, Prossnitz ER, Sklar LA (2004) Real-time analysis of ternary complex on particles: direct evidence for partial agonism at the agonist-receptor-G protein complex assembly step of signal transduction. J Biol Chem 279:13514–13521

Sklar LA, Oades ZG, Jesaitis AJ, Painter RG, Cochrane CG (1981) Fluoresceinated chemotactic peptide and high-affinity antifluorescein antibody as a probe of the temporal characteristics of neutrophil stimulation. Proc Natl Acad Sci USA 78:7540–7544

Sklar LA, Finney DA, Oades ZG, Jesaitis AJ, Painter RG, Cochrane CG (1984) The dynamics of ligand-receptor interactions. Real-time analyses of association, dissociation, and internalization of an N-formyl peptide and its receptors on the human neutrophil. J Biol Chem 259:5661–5669

Sklar LA, Hyslop PA, Oades ZG, Omann GM, Jesaitis AJ, Painter RG, Cochrane CG (1985) Signal transduction and ligand-receptor dynamics in the human neutrophil. Transient responses and occupancy–response relations at the formyl peptide receptor. J Biol Chem 260:11461–11467

Sklar LA, Bokoch GM, Button D, Smolen JE (1987) Regulation of ligand-receptor dynamics by guanine nucleotides. Real-time analysis of interconverting states for the neutrophil formyl peptide receptor. J Biol Chem 262:135–139

Sklar LA, Mueller H, Omann G, Oades Z (1989) Three states for the formyl peptide receptor on intact cells. J Biol Chem 264:8483–8486

Sklar LA, Vilven J, Lynam E, Neldon D, Bennett TA, Prossnitz E (2000) Solubilization and display of G protein-coupled receptors on beads for real-time fluorescence and flow cytometric analysis. Biotechniques 28:976–985

Sklar LA, Edwards BS, Graves SW, Nolan JP, Prossnitz ER (2002) Flow cytometric analysis of ligand-receptor interactions and molecular assemblies. Annu Rev Biophys Biomol Struct 31:97–119

Vauquelin G, Van Liefde I (2005) G protein-coupled receptors: A count of 1001 conformations. Fundam Clin Pharmacol 19:45–56

Vilven JC, Domalewski M, Prossnitz ER, Ye RD, Muthukumaraswamy N, Harris RB, Freer RJ, Sklar LA (1998) Strategies for positioning fluorescent probes and crosslinkers on formyl peptide ligands. J Recept Signal Transduct Res 18:187–221

Wise A, Gearing K, Rees S (2002) Target validation of G-protein coupled receptors. Drug Discov Today 7:235–246

Ernst Schering Foundation Symposium Proceedings, Vol. 2, pp. 229–247
DOI 10.1007/2789_2006_011

Published Online: 16 May 2007

High Content Screening to Monitor G Protein-Coupled Receptor Internalisation

R. Heilker(✉)

Boehringer Ingelheim Pharma GmbH Co. KG, Department of Lead Discovery, Birkendorfer Strasse 65, 88397 Biberach an der Riss, Germany
email: *Ralf.Heilker@bc.boehringer-ingelheim.com*

Abstract. G protein-coupled receptors (GPCRs) fulfil a broad diversity of physiological functions in areas such as neurotransmission, respiration, cardiovascular action, pain and more. Consequently, they are considered as the most successful group of therapeutic targets on the pharmaceutical market, and the search for compounds that interfere with GPCR function in a specific and selective way is a major focus of the pharmaceutical industry. High Content Screening (HCS), a combination of fluorescence microscopic imaging and automated image analysis, has become a frequently employed tool to study test compound effects in cellular disease modelling systems. One way to functionally analyse the effect of test compounds on GPCRs by HCS relies on the broadly observed phenomenon of desensitisation. Agonist stimulation of most GPCRs leads to their intracellular phosphorylation and subsequent internalisation, resulting in the termination of receptor signalling and the seclusion of the GPCR from further extracellular stimulation. Complementary to other functional GPCR drug discovery assays, GPCR internalisation assays enable a desensitisation-focussed pharmacological analysis of test compounds.

1 High Content Screening

Fluorescence microscopy has been widely employed in academic cell biology research as a non-destructive and sensitive technique to visualise subcellular structures and to monitor intracellular protein translocations. In recent years, the pharmaceutical industry's interest in studying test compounds in cellular assays has continuously increased. In particular, a novel technique generally referred to as High Content Screening (HCS) has been introduced, which combines high-resolution fluorescence microscopy with automated image analysis (Conway et al. 1999; Ghosh et al. 2000; Taylor et al. 2001; Li et al. 2003; Almholt et al. 2004). After fluorophore labelling on a cellular level, the biomolecules of interest are observed by fluorescence microscopy, possibly in parallel at different wavelengths (multiplexing). Appropriate image analysis algorithms then quantify the distribution and brightness of the fluorophore-labelled biomolecules in the cells. In live cell experiments, the kinetics of a pharmaceutical drug effect and respective intracellular protein trafficking events can be monitored. Apart from protein trafficking (Almholt et al. 2004), HCS can provide information on the phosphorylation state of target proteins (Russello 2004), on cellular proliferation (Bhawe et al. 2004) or apoptosis (Steff et al. 2001), on morphological changes such as neurite outgrowth (Simpson et al. 2001), on modifications of the cytoskeleton (Olson and Olmsted 1999; Giuliano 2003), on cellular movements (Soll et al. 2000) and other overall changes of the fluorescence such as for the analysis of gap junctions (Li et al. 2003).

By this means, HCS provides several advantages over standard high-throughput screening (HTS). Cellular HTS conventionally examines the mean response of the whole cell population in a microtitre plate (MTP) well. In contrast, HCS distinguishes the individual response of many cells in an MTP well. The individual cells may differ with respect to the differentiation, the stage of the cell cycle, the state of transfection or because of natural variability. In consequence, heterogeneous pharmaceutical drug effects on mixed cell populations may be analysed in a single MTP well. On-target drug effects (meaning pharmaceutical effects of a test compound that are directly related to the targeted biomolecule of interest) may be cross-correlated with other phenomena such as cellular toxicity (Wolff et al. 2005). Compound artefacts such as cell lysis or compound autofluorescence are readily discovered. HCS permits work with recombinantly or endogenously expressed biomolecules of interest in cell lines or primary cells given that suitable detection reagents such as specific antibodies are available. In summary, HCS enables completely novel assay formats, which do not rely on an overall change of fluorescence or luminescence intensity from the whole MTP well. The development of high-throughput fluorescence microscopic imaging devices and of rapid automated image analysis algorithms has enabled the drug discovery application of the HCS format to high test compound numbers.

2 Confocal Optics for HCS

In a standard widefield microscopic image, light is also collected from outside the focal plane. Imaging of intracellular structures has benefited substantially from the introduction of confocal microscopy. Fundamentally, confocal optics dramatically improve the spatial resolution in the vertical direction, greatly reducing interference from adjacent object features above or below the focal plane. In the microscopic analysis of a cell population adherent to the bottom of an MTP well, confocal optics enable the observation of the cell layer in the focal plane without interference from dead cells, free fluorophores or autofluorescent particles above the cellular layer. This increased optical resolution improves the visualisation of the complex subcellular membrane, vesicle and organelle systems within eukaryotic cells. The detailed study of intracellular translocation of target biomolecules, for example the translocation of a transcription factor from the cytosol to the nucleus in response to a stimulus, is facilitated by this approach.

Several systems are commercially available for confocal microscopic imaging of cells.

In classic confocal optics (Wilson 1990), a high numerical aperture objective lens is employed to focus the excitation laser light to the focal plane of interest. The restriction to fluorescence emission from a particular confocal observation volume is achieved by guiding the emitted light through a pinhole (Fig. 1a). To obtain high-fluorescence sensitivity, low-noise detectors such as avalanche photodiodes are employed. Using the confocal detection principle one can analyse a femtolitre-sized sample volume, largely free of background from solvent Rayleigh and Raman scattering as well as from fluorescent impurities. The confocal observation volume can be scanned through living cells for the study of fluorescent biomolecules in their native environments.

However, the available confocal point scanning microscopes, which are based on the single-pinhole principle, are generally too slow for drug

Fig. 1 a–d. a Confocal optics. For confocal fluorescence studies, a high numerical aperture objective lens focuses the excitation light (*solid lines*) into the sample. The emission light (*dotted lines*) is collected through the same objective lens and guided through a pinhole to restrict fluorescence detection to a femtolitre-sized confocal observation volume in the sample. **b** Nipkow disk. *Left*, point scanning: the laser focus is moved in a raster scanning mode across the specimen. A single pinhole in the emission light path restricts the detection to the focal plane. *Right*, Nipkow disk: a multitude of pinholes in the so-called Nipkow disk creates a series of confocal sample volumes that probe the specimen in parallel (the scheme is a kind gift of Dr. Phil Vanek; Becton, Dickinson). **c** Schematic light path of the Opera™. The excitation laser light is guided through a rotating microlens array that focuses the light through a dichroic mirror, a simultaneously rotating pinhole array and an objective lens to the sample. Fluorescence emission from the sample is guided back via the objective and pinhole array, then reflected towards a CCD camera (scheme is a kind gift of Dr. Gabriele Gradl; Evotec Technologies). **d** Schematic light path of the IN Cell Analyzer 3000™. The excitation laser light is autofocused through an objective to the bottom of the microtitre plate. Fluorescence emission is collected by the same objective, then guided through a dichroic mirror and confocal slit mask to three simultaneously operating CCD detectors for different wavelength detection (scheme is a kind gift of Dr. Gerd Erhard; General Electrics Healthcare Biosciences)

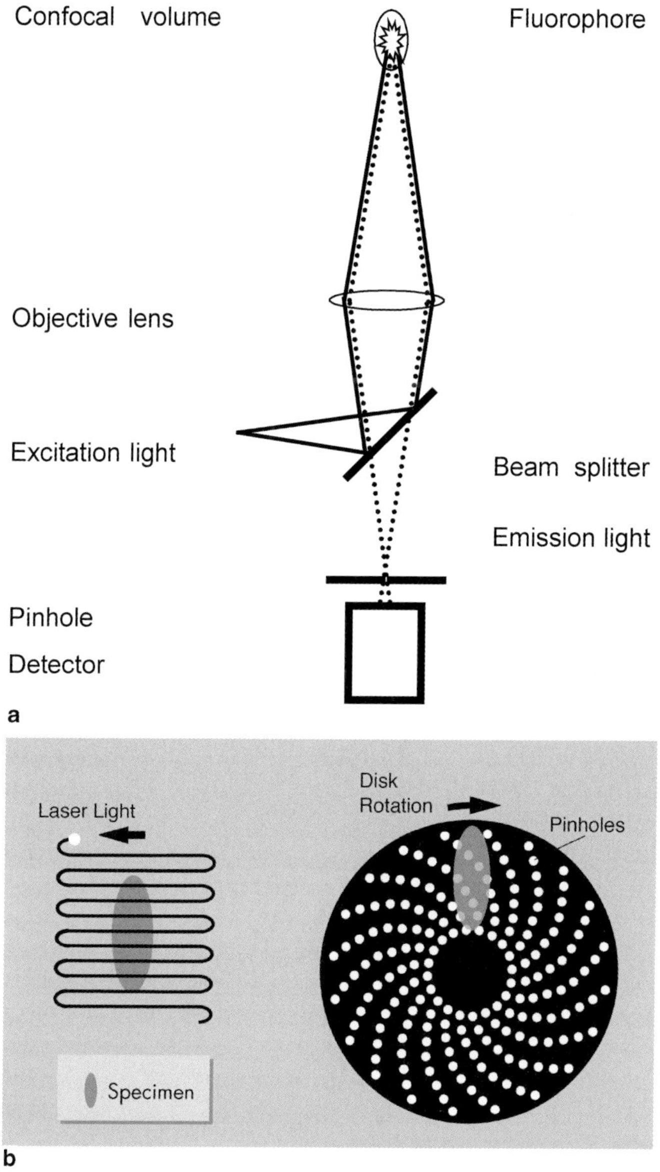
Confocal volume
Fluorophore
Objective lens
Excitation light
Beam splitter
Emission light
Pinhole
Detector
a
Laser Light
Disk
Rotation
Pinholes
Specimen
b

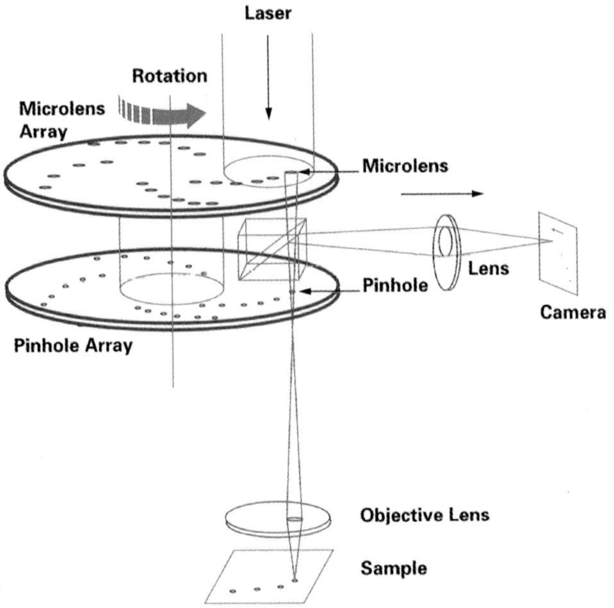
Laser
Rotation
Microlens Array
Microlens
Lens
Pinhole
Camera
Pinhole Array
Objective Lens
Sample

c

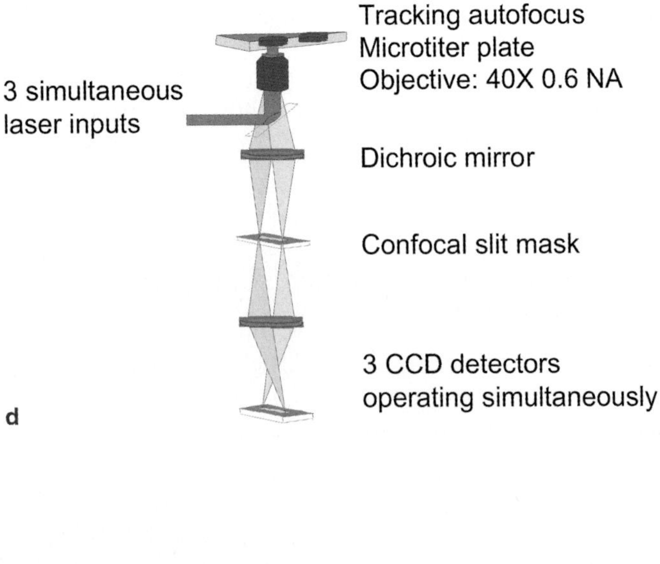
Tracking autofocus
Microtiter plate
Objective: 40X 0.6 NA
3 simultaneous laser inputs
Dichroic mirror
Confocal slit mask
3 CCD detectors operating simultaneously

d

screening applications. The first three confocal high-throughput cellular imagers (Zemanova et al. 2003) marketed to fill this gap were the Opera™ from Evotec Technologies GmbH (Hamburg, Germany), the IN Cell Analyzer 3000™ from General Electric Healthcare Biosciences (Amersham, United Kingdom) and the BD Pathway Bioimager from Becton, Dickinson and Company (San Jose, CA, USA). These systems achieve a readout time of down to approximately 1 s per well (varying for example with the required resolution of the microscopic image and the brightness of the fluorophores). For the sake of higher throughput, the Opera™ and the BD Pathway Bioimager™ employ a Nipkow disk (Fig. 1b,c) to project fluorescence from several confocal volumes in parallel to a CCD camera. In a similar approach to shorten the imaging time, the IN Cell Analyzer 3000™ employs line scanning through a confocal slit (Fig. 1d). This new generation of HTS-compatible confocal imaging readers combines high-temporal with high-spatial resolution. To further increase throughput, all three systems support an autofocus mechanism that keeps the microscope objective focused to the cellular layer adherent to the bottom of the well. For live cell measurements, the Opera™, BD Pathway Bioimager™ and IN Cell Analyzer 3000™ provide environmental chambers, which maintain user-defined temperature and carbon dioxide levels.

The Opera™ and the IN Cell Analyzer 3000™ employ four or respectively three laser sources and four or respectively three CCD cameras for synchronous, multiplexed imaging, adding to a potential increase of throughput.

In a direct comparison between a confocal and a non-confocal HCS imaging device, the confocal optics provided a better basis for the detection and quantification of smaller subcellular structure (Haasen et al. 2006b).

3 Pharmaceutical Analysis of GPCR Ligands: Binding Versus Function

GPCRs possess an N-terminal extracellular domain and a C-terminal intracellular domain. They are held in a lipid bilayer by seven membrane-spanning helices (Ji et al. 1998), connected by three extracellular and three intracellular loops. Specific agonist binding to the GPCR causes a conformational rearrangement of the receptor transmembrane helices, which leads to the binding and activation of an intracellular heterotrimeric G-protein (Perez and Karnik 2005). By these means, GPCRs act as sensors of exogenous signals, which they transduce into cytoplasmic signalling pathways. The first GPCRs to be cloned were bovine opsin (Nathans and Hogness 1983) and the beta-adrenergic receptor (Dixon et al. 1986). Since then, a large gene family of a further approximately 2000 GPCRs has been reported, classified into more than 100 subfamilies according to sequence homology, ligand structure, and receptor function.

GPCRs are the most important class of therapeutic targets (Ma and Zemmel 2002). Approximately 45% of all known pharmaceutical drugs are directed against transmembrane receptors (Drews 2000), largely against GPCRs. As GPCRs are involved in a broad diversity of physiological functions, the modulation of GPCR signalling is a major focus of pharmaceutical research.

In the past, the interaction between GPCRs and their extracellular ligands has proven to be an attractive point of interference for therapeutic agents. For that reason, the pharmaceutical industry has developed numerous biochemical drug discovery assays to investigate these ligand–GPCR interactions, such as the scintillation proximity assays (SPA) (Alouani 2000) or the less frequently employed fluorescence polarisation (FP) assays (Banks and Harvey 2002; Harris et al. 2003) and fluorescence intensity distribution analysis (FIDA) assays (Auer et al. 1998; Zemanova et al. 2003). All the above-mentioned biochemical binding assays (Heilker et al. 2005) are based upon the competition of the test compound with a labelled reference ligand. An obvious disadvantage of these reference ligand displacement assays is the risk of missing non-competitive, allosteric ligands. Further, the binding assay does not reveal the functional activity of the test compounds as full/partial agonism, neutral antagonism, inverse agonism or positive modulation. To supplement compound testing in this direction, there is a need for functional assay formats, possibly measuring GPCR activity in a more physiological, cellular background.

GPCR signal transduction mechanisms have been categorised into three major classes, based upon the Gα subunit employed in the initial signalling step: $G\alpha_q$ (phospholipase C), $G\alpha_i$ and $G\alpha_s$ (inhibition and stimulation of cAMP production, respectively).

The most broadly applied cell-based technique to measure $G\alpha_q$-mediated signalling is the Ca release assay, either measured in a fluorescent format using Ca-sensitive fluorophores (Sullivan et al. 1999) or in a luminescent format using aequorin and a chemiluminescent substrate (Dupriez et al. 2002). $G\alpha_i$- and $G\alpha_s$-mediated modulation of the cytosolic cyclic adenosine monophosphate (cAMP) content may be analysed using various detection technologies (Gabriel et al. 2003).

Alternatively to measuring the cellular signalling via G-proteins, the functional activation of GPCRs may be investigated by agonist-induced receptor internalisation (Milligan 2003). The broad applicability of GPCR internalisation assays (Fig. 2; scheme of GPCR internalisation) is based on the common phenomenon of GPCR desensitisation and has been demonstrated for numerous GPCRs (Krupnick and Benovic 1998; Ferguson 2001; Oakley et al. 2002). In the desensitisation process, GPCR kinases (GRKs) phosphorylate agonist-activated GPCRs on serine and threonine residues. Arrestins are cytoplasmic proteins that are recruited to the plasma membrane by GRK-phosphorylated GPCRs (Barak et al. 1997). Arrestins then uncouple the GPCR from the cognate G protein (Lohse et al. 1992; Pippig et al. 1993) and target the desensitised receptors to clathrin-coated pits for endocytosis (Goodman et al. 1996; Laporte et al. 2000).

Fig. 2a–d. Scheme of GPCR internalisation assays. After agonist [A] stimulation, the GPCR first recruits a heterotrimeric G-protein ($\alpha\beta\gamma$), which is activated and released. The GPCR then becomes phosphorylated by a GRK on its carboxy-terminal tail. β-Arrestin (β-AR) is recruited to the plasma membrane by the GRK-phosphorylated GPCR. β-AR then targets the GPCR to clathrin-coated pits for endocytosis. Depending on the stability of the specific GPCR–arrestin interaction, β-AR is either released after the formation of clathrin-coated pits or co-internalised with the GPCR-loaded vesicles. The internalisation process may be monitored by **a** a GFP-fusion with β-AR, **b** a fluorophore [F] label on the agonist, **c** a fluorophore-conjugated antibody to label the GPCR or **d** a GFP fusion with the GPCR

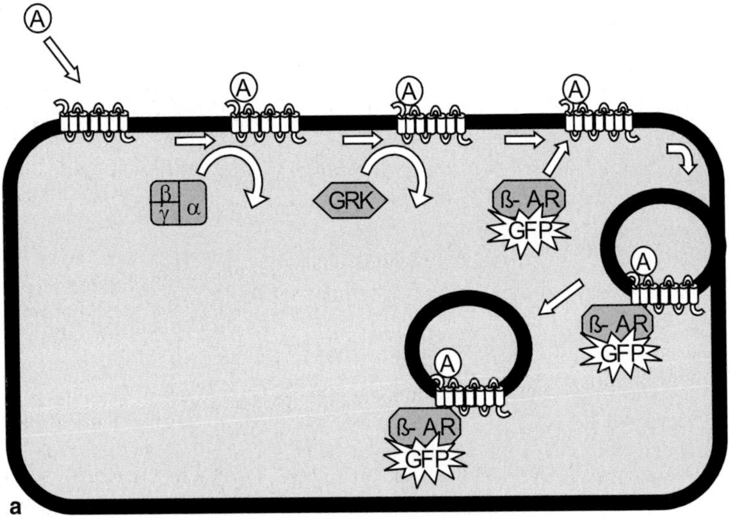
A
A
A
A
β
γ
α
GRK
ß- AR
GFP
A
ß- AR
GFP
A
ß- AR
GFP

a

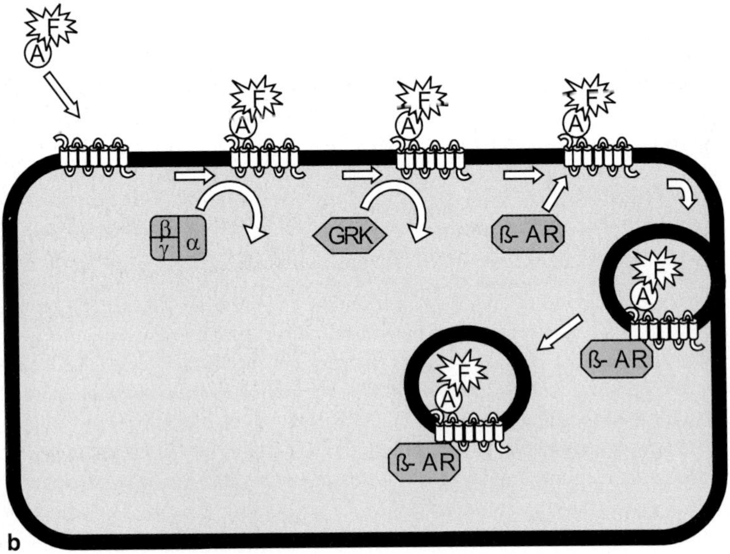
F
A
F
A
F
A
F
A
β
γ
α
GRK
ß- AR
F
A
ß- AR
F
A
ß- AR

b

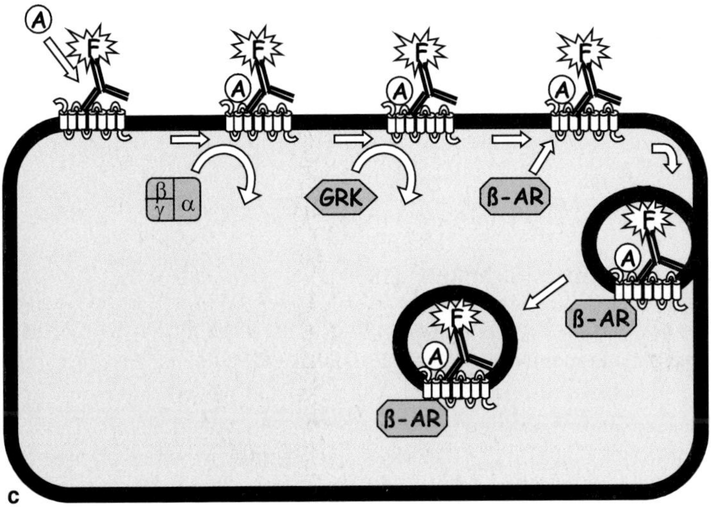
A
F
F
A
F
A
F
A
β
γ
α
GRK
ß-AR
F
A
ß-AR
F
A
ß-AR

c

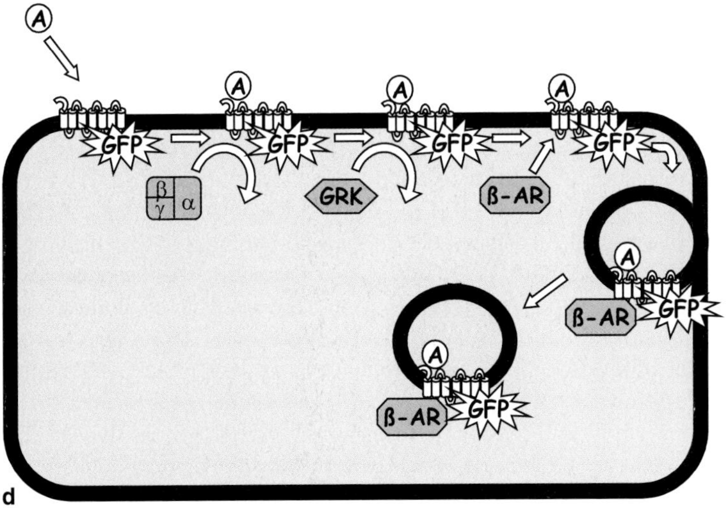
A
A
A
A
GFP
GFP
GFP
GFP
β
γ
α
GRK
ß-AR
A
ß-AR
GFP
A
ß-AR
GFP

d

In contrast to Ca release and cAMP assays, desensitisation occurs independently of the associated G-protein subclass. Further, the imaging-based GPCR internalisation assays offer the general advantages of the HCS format as described above.

4 GPCR Internalisation Assays for HCS

4.1 Transfluor Technique

As described above, arrestins are cytoplasmic proteins that are recruited to the plasma membrane by ligand-activated GPCRs and then target the GRK-desensitised receptors to clathrin-coated pits for internalisation. The Transfluor™ technology, a licensed product of Molecular Devices (Sunnyvale, CA, USA), is an HCS assay format in which an arrestin-green fluorescent protein (ArrGFP) conjugate redistributes upon GPCR stimulation (Fig. 2a; Oakley et al. 2002; Milligan 2003). As most GPCRs undergo desensitisation, the translocation of ArrGFP has been observed for numerous GPCRs independent of the interacting G-protein and of the class of ligand bound by the receptor (Oakley et al. 2002).

If the interaction between arrestin and the GPCR of interest is of low affinity, e.g. for the β2 adrenergic receptor, the ArrGFP is released after the formation of clathrin-coated pits (Oakley et al. 2000, 2001). If the interaction between ArrGFP and the GPCR is of high affinity, e.g. for the vasopressin 2 receptor, the arrestin is co-internalised with the GPCR to endocytic vesicles. Respectively, two Transfluor™ object types may be observed in fluorescence microscopy: the smaller, less fluorescently bright ArrGFP-labelled coated pits formed by the GPCRs with low arrestin affinity and the larger, more fluorescently bright ArrGFP-labelled endocytic vesicles formed by the GPCRs with higher arrestin affinity. Earlier work with these two Transfluor™ receptor types showed that both confocal and non-confocal HCS imaging devices can optically resolve the fluorophore-labelled coated pits and endocytic vesicles to enable image analysis (Oakley et al. 2002). Particularly for the GPCRs that carry the ArrGFP only to the coated pit level, however, confocal optics provided better assay statistics in a direct comparison to a non-confocal imaging device (Haasen et al. 2006b).

4.2 Labelled Ligand Internalisation

Another means to monitor the internalisation of a GPCR is to co-internalise a specific, fluorophore-labelled ligand (Fig. 2b; Haasen et al. 2006a). A challenge herein is the possible interference of the fluorophore label with ligand binding to the receptor. The labelled ligand protocol for GPCR internalisation assays makes it possible to distinguish between orthosteric and allosteric GPCR-binding test compounds. Orthosterically acting test compounds block the binding of the labelled reference ligand to the cell surface and cannot further be analysed in labelled ligand-assays. An allosterically acting test compound permits the binding of the labelled reference ligand to the plasma membrane-exposed GPCRs. If the mechanism of action is antagonistic, a pre-incubation of the allosteric test compound prevents the co-internalisation of the labelled reference ligand. If the mechanism is positive modulatory, a pre-incubation of the allosteric test compound leads to an increased rate of labelled-ligand internalisation. With regard to agonistic, allosteric test compounds, the interpretation of the experimental results becomes more challenging: if added at the same time as the labelled reference ligand, depending on the ratio of kinetics, potencies and efficacies between test compound and reference ligand, the test compounds may either increase or decrease the labelled-ligand internalisation rate.

4.3 Labelled Receptor Internalisation

An alternative assay technology to follow the internalisation of a GPCR is to co-internalise a specific fluorophore-labelled antibody, either directed against an extracellular domain of the receptor, or against an amino-terminal epitope tag (Fig. 2c; Haasen et al. 2006a). Test compounds can be qualified as agonists or antagonists in these labelled-GPCR internalisation experiments. After pre-incubation of the GPCR-overexpressing cells with a GPCR-specific antibody, the test compounds may be analysed for their internalisation-agonistic properties. Alternatively, if the test compounds are pre-incubated with the GPCR-expressing cells, the non-occurrence of reference agonist-induced GPCR internalisation indicates antagonistic/inverse agonistic properties of the test compound. If the constitutive internalisation and recycling rates of the GPCR are adequate, inverse agonists differ from the neutral antagonists by inducing an enrichment of the receptor at the plasma membrane. In a further experimental variation, the assay conditions can be adapted to analyse positive modulators: if the reference agonist is added at approximately EC10 concentration, a pre-incubation with a positive modulator compound will significantly increase the internalisation rate.

A challenge in GPCR labelling assays is that antibodies against the extracellular portion of the native receptor are not always available. Further, it is important that the GPCR-labelling antibody does not interfere with ligand binding and does not produce a functional effect itself. To reduce the probability of such issues, an N-terminal tag may be attached to the GPCR of interest, so that a tag-directed antibody can be used for GPCR detection.

If no antibody against an extracellular GPCR epitope is available for co-internalisation experiments, the assay protocol may be modified so that the GPCR of interest is antibody-labelled *after* completion of the internalisation step. In this case, labelling is conducted in the paraformaldehyde-fixed and detergent-solubilised cells so that the antibody may also be directed against an intracellular epitope of the GPCR.

Another way to label the GPCR of interest is the use of a fluorescent protein (FP) label, typically at the C-terminus of the receptor (Fig. 2d; Xia et al. 2004). A possible risk of this approach is that the FP fusion can change the receptor pharmacology or interfere with the induction of intracellular signalling. Similar to the post-fixation labelling of the GPCR with an antibody described above, such an FP label monitors the ligand-induced net shift of local GPCR concentrations between the plasma membrane and intracellular compartments.

4.4 GPCR Internalisation Assays: Synopsis

The Transfluor™ technique is the most frequently described format for a GPCR internalisation assay. One advantage of Transfluor™ is that it does not require a modification of the reference ligand or of the receptor interest. Test compounds can be analysed for their interference with the unaltered receptor ligand pair. The assay format has been described as very robust in numerous publications.

Both the labelled reference ligand internalisation and the labelled GPCR internalisation methods can, however, also produce stable HCS assays. Both formats provide advantageous, partially complementary features in the characterisation of GPCR ligand test compounds. The fluorophore label on the GPCR enables the analysis of both orthosteric and allosteric test compounds. A possible disadvantage of this technique is the modification of the receptor pharmacology by the label. The ligand label protocol is particularly well suited to investigate allosteric GPCR-binding test compounds: these allosteric compounds do not displace the fluorophore-labelled reference ligands and therefore enable the monitoring of both internalisation pathway and kinetics.

For internalisation assays with an extracellular ligand/GPCR labelling step at the plasma membrane, the initially increasing number and intensity of intracellular fluorescent granules describes the rate of receptor internalisation. After a few minutes, however, (a) the newly synthesised protein from the secretory pathway reaching the cell surface, (b) recycling and/or (c) degradation of ligand- or antibody-labelled GPCRs will be superimposed on the internalisation kinetics. To help unravel these complex intracellular receptor trafficking pathways, either the FP label at the C-terminus of the GPCR or the use of a GPCR-detecting antibody on fixed and permeabilised cells can serve as a second label to describe the overall distribution of the GPCR. Such a second label enables the observation of the ligand-induced net shift of local GPCR concentrations between the plasma membrane and intracellular compartments. Further, such labelling provides an impression of the GPCR distribution in the cells prior to ligand stimulation and thereby facilitates the overall investigation of the test compound-induced changes.

The option to use fixed-cell internalisation endpoint protocols is advantageous for automated liquid handling and offline imaging. Thus, receptor internalisation assays can provide robust medium and/or high-throughput screening (MTS/HTS) formats and complement the drug discovery tool spectrum for the GPCR target class very well.

References

Almholt DL, Loechel F, Nielsen SJ, Krog-Jensen C, Terry R, Bjorn SP, Pedersen HC, Praestegaard M, Moller S, Heide M, Pagliaro L, Mason AJ, Butcher S, Dahl SW (2004) Nuclear export inhibitors and kinase inhibitors identified using a MAPK-activated protein kinase 2 redistribution screen. Assay Drug Dev Technol 2:7–20

Alouani S (2000) Scintillation proximity binding assay. Methods Mol Biol 138:135–141

Auer M, Moore KJ, Meyer-Almes FJ, Guenther R, Pope AJ, Stoeckli K (1998) Fluorescence correlation spectroscopy: lead discovery by miniaturized HTS. Drug Discov Today 3:457–465

Banks P, Harvey M (2002) Considerations for using fluorescence polarization in the screening of g protein-coupled receptors. J Biomol Screen 7:111–117

Barak LS, Ferguson SS, Zhang J, Caron MG (1997) A beta-arrestin/green fluorescent protein biosensor for detecting G protein-coupled receptor activation. J Biol Chem 272:27497–27500

Bhawe KM, Blake RA, Clary DO, Flanagan PM (2004) An automated image capture and quantitation approach to identify proteins affecting tumor cell proliferation. J Biomol Screen 9:216–222

Conway BR, Minor LK, Xu JZ, Gunnet JW, DeBiasio R, D'Andrea MR, Rubin R, DeBiasio R, Giuliano K, DeBiasio L, Demarest KT (1999) Quantification of G-protein coupled receptor internatilization using G-protein coupled receptor-green fluorescent protein conjugates with the ArrayScan high-content screening system. J Biomol Screen 4:75–86

Dixon RA, Kobilka BK, Strader DJ, Benovic JL, Dohlman HG, Frielle T, Bolanowski MA, Bennett CD, Rands E, Diehl RE (1986) Cloning of the gene and cDNA for mammalian beta-adrenergic receptor and homology with rhodopsin. Nature 321:75–79

Drews J (2000) Drug discovery: a historical perspective. Science 287:1960–1964

Dupriez VJ, Maes K, Le Poul E, Burgeon E, Detheux M (2002) Aequorin-based functional assays for G-protein-coupled receptors, ion channels, and tyrosine kinase receptors. Receptors Channels 8:319–330

Ferguson SS (2001) Evolving concepts in G protein-coupled receptor endocytosis: the role in receptor desensitization and signaling. Pharmacol Rev 53:1–24

Gabriel D, Vernier M, Pfeifer MJ, Dasen B, Tenaillon L, Bouhelal R (2003) High throughput screening technologies for direct cyclic AMP measurement. Assay Drug Dev Technol 1:291–303

Ghosh RN, Chen YT, DeBiasio R, DeBiasio RL, Conway BR, Minor LK, Demarest KT (2000) Cell-based, high-content screen for receptor internalization, recycling and intracellular trafficking. Biotechniques 29:170–175

Giuliano KA (2003) High-content profiling of drug–drug interactions: cellular targets involved in the modulation of microtubule drug action by the antifungal ketoconazole. J Biomol Screen 8:125–135

Goodman OB Jr, Krupnick JG, Santini F, Gurevich VV, Penn RB, Gagnon AW, Keen JH, Benovic JL (1996) Beta-arrestin acts as a clathrin adaptor in endocytosis of the beta2-adrenergic receptor. Nature 383:447–450

Haasen D, Schnapp A, Valler MJ, Heilker R (2006a) G protein-coupled receptor internalization assays in the high content screening format. Methods Enzymol 414:121–139

Haasen D, Wolff M, Valler MJ, Heilker R (2006b) Comparison of G-protein coupled receptor desensitization-related beta-arrestin redistribution using confocal and non-confocal imaging. Comb Chem High Throughput Screen 9:37–47

Harris A, Cox S, Burns D, Norey C (2003) Miniaturization of fluorescence polarization receptor-binding assays using CyDye-labeled ligands. J Biomol Screen 8:410–420

Heilker R, Zemanova L, Valler MJ, Nienhaus GU (2005) Confocal fluorescence microscopy for high-throughput screening of G-protein coupled receptors. Curr Med Chem 12:2551–2559

Ji TH, Grossmann M, Ji I (1998) G protein-coupled receptors. I. Diversity of receptor-ligand interactions. J Biol Chem 273:17299–17302

Krupnick JG, Benovic JL (1998) The role of receptor kinases and arrestins in G protein-coupled receptor regulation. Annu Rev Pharmacol Toxicol 38:289–319

Laporte SA, Oakley RH, Holt JA, Barak LS, Caron MG (2000) The interaction of beta-arrestin with the AP-2 adaptor is required for the clustering of beta 2-adrenergic receptor into clathrin-coated pits. J Biol Chem 275:23120–23126

Li Z, Yan Y, Powers EA, Ying X, Janjua K, Garyantes T, Baron B (2003) Identification of gap junction blockers using automated fluorescence microscopy imaging. J Biomol Screen 8:489–499

Lohse MJ, Andexinger S, Pitcher J, Trukawinski S, Codina J, Faure JP, Caron MG, Lefkowitz RJ (1992) Receptor-specific desensitization with purified proteins. Kinase dependence and receptor specificity of beta-arrestin and arrestin in the beta 2-adrenergic receptor and rhodopsin systems. J Biol Chem 267:8558–8564

Ma P, Zemmel R (2002) Value of novelty? Nat Rev Drug Discov 1:571–572

Milligan G (2003) High-content assays for ligand regulation of G-protein-coupled receptors. Drug Discov Today 8:579–585
Nathans J, Hogness DS (1983) Isolation, sequence analysis, and intron-exon arrangement of the gene encoding bovine rhodopsin. Cell 34:807–814
Oakley RH, Laporte SA, Holt JA, Caron MG, Barak LS (2000) Differential affinities of visual arrestin, beta arrestin1, and beta arrestin2 for G protein-coupled receptors delineate two major classes of receptors. J Biol Chem 275:17201–17210
Oakley RH, Laporte SA, Holt JA, Barak LS, Caron MG (2001) Molecular determinants underlying the formation of stable intracellular G protein-coupled receptor-beta-arrestin complexes after receptor endocytosis*. J Biol Chem 276:19452–19460
Oakley RH, Hudson CC, Cruickshank RD, Meyers DM, Payne RE, Rhem SM, Loomis CR (2002) The cellular distribution of fluorescently labeled arrestins provides a robust, sensitive, and universal assay for screening of G protein-coupled receptors. Assay Drug Dev Technol 1:21–30
Olson KR, Olmsted JB (1999) Analysis of microtubule organization and dynamics in living cells using green fluorescent protein-microtubule-associated protein 4 chimeras. Methods Enzymol 302:103–120
Perez DM, Karnik SS (2005) Multiple signaling states of G-protein-coupled receptors. Pharmacol Rev 57:147–161
Pippig S, Andexinger S, Daniel K, Puzicha M, Caron MG, Lefkowitz RJ, Lohse MJ (1993) Overexpression of beta-arrestin and beta-adrenergic receptor kinase augment desensitization of beta 2-adrenergic receptors. J Biol Chem 268:3201–3208
Russello SV (2004) Assessing cellular protein phosphorylation: high throughput drug discovery technologies. Assay Drug Dev Technol 2:225–235
Simpson PB, Bacha JI, Palfreyman EL, Woollacott AJ, McKernan RM, Kerby J (2001) Retinoic acid evoked-differentiation of neuroblastoma cells predominates over growth factor stimulation: an automated image capture and quantitation approach to neuritogenesis. Anal Biochem 298:163–169
Soll DR, Voss E, Johnson O, Wessels D (2000) Three-dimensional reconstruction and motion analysis of living, crawling cells. Scanning 22:249–257
Steff AM, Fortin M, Arguin C, Hugo P (2001) Detection of a decrease in green fluorescent protein fluorescence for the monitoring of cell death: an assay amenable to high-throughput screening technologies. Cytometry 45:237–243
Sullivan E, Tucker EM, Dale IL (1999) Measurement of [Ca2+] using the Fluorometric Imaging Plate Reader (FLIPR). Methods Mol Biol 114:125–133
Taylor DL, Woo ES, Giuliano KA (2001) Real-time molecular and cellular analysis: the new frontier of drug discovery. Curr Opin Biotechnol 12:75–81

Wilson T (1990) Confocal microscopy. Academic Press, London
Wolff M, Haasen D, Merk S, Kroner M, Maier U, Bordel S, Wiedenmann J, Nienhaus GU, Valler MJ, Heilker R (2005) Automated high content screening for phosphoinositide kinase 3 inhibition using an AKT1 redistribution assay. Comb Chem High Throughput Screen 9:339–350
Xia S, Kjaer S, Zheng K, Hu PS, Bai L, Jia JY, Rigler R, Pramanik A, Xu T, Hokfelt T, Xu ZQ (2004) Visualization of a functionally enhanced GFP-tagged galanin R2 receptor in PC12 cells: constitutive and ligand-induced internalization. Proc Natl Acad Sci USA 101:15207–15212
Zemanova L, Schenk A, Valler MJ, Nienhaus GU, Heilker R (2003) Confocal optics microscopy for biochemical and cellular high-throughput screening. Drug Discov Today 8:1085–1093

Ernst Schering Foundation Symposium Proceedings, Vol. 2, pp. 249–267
DOI 10.1007/2789_2006_012

Published Online: 16 May 2007

High-Throughput Lead Finding and Optimisation for GPCR Targets

A. Sewing(✉), D. Cawkill

Primary Pharmacology Group, Pfizer PDGRD, IPC 580, Ramsgate Road, CT13 9NJ Sandwich, UK
email: *Andreas.Sewing@pfizer.com*

Abstract. Driven by past successes and the detailed knowledge of signalling cascades and physiological processes, G-protein-coupled receptors are taking a prominent place in the portfolios of many pharmaceutical companies. To successfully address this target class, scientists need not only a good understanding of the specific receptor under investigation, but also the right tools from assay technology, reagent production to a hit-to-lead process that acknowledges the importance of parameters beyond potency and embraces the gain in knowledge of the last decade. This manuscripts attempts to summarise some of the changes and progress made across the pharmaceutical industry to design an efficient and effective strategy for finding and optimising small molecules modulating the activity of GPCRs.

1 Introduction

Altering physiological or disease-related processes through modulation of G-protein-coupled receptors (GPCRs) and their associated signalling cascades is one of the most successful strategies in modern drug discovery. It has produced a number of blockbuster drugs and today GPCRs account for approximately 30% of small molecule drug targets (Hopkins and Groom 2002). Looking at the 100 top-selling drugs, 25% target GPCRs and these contributed over $30 billion to annual sales in 2001 (Klabunde and Hessler 2002). Given these numbers coupled with the good accessibility of transmembrane receptors, tissue-specific expression, and their predicted druggability (i.e., the ability to bind a small rule of 5 compliant organic molecule (Hopkins and Groom 2002)), it is not surprising that GPCRs remain one of the most important target classes and feature prominently in the drug discovery portfolio of pharmaceutical companies. From the analysis of the sequenced human genome, the predicted overall number of GPCRs is in the region of 865 (Fredriksson and Schiöth 2005). With several hundred members of the sensory type and approximately 210 GPCRs with known endogenous ligands (Fredriksson and Schiöth 2005) (only a portion are pursued as drug targets), this still leaves a sizable number of so-called orphan GPCRs (i.e., with no known ligands), representing a potential pool of new targets for pharmaceutical companies.

Even with the historic success of GPCR-directed drug discovery, there are a number of challenges and questions that require closer examination: the underlying hypothesis of scientists and analysts alike is that the success rates can be repeated with new targets. However, as with most other targets, drugs were historically discovered serendipitously and success rates, with new paradigms applied in all areas of research, remain to be established. Indeed, in the area of high-throughput screening the perception in the literature is that reality has been lagging behind expectations (Hefti and Bolten 2003; Peakman et al. 2003; Sewing and Gribbon 2005). With orphan GPCR receptors, we have seen the rapid advance of knowledge and the de-orphanisation of a number of receptors, yet it remains to be established how much value can be extracted and progress has been slow in many cases (Nambi and Aiyar 2003; Glasel 2004). Looking at the GPCRs that have been successfully addressed with small organic molecules, a large proportion target receptors with aminergic ligands with less success coming from peptide-ligand GPCRs, a fact attributed to the potentially unfavourable physicochemical profile of molecules targeting peptidic GPCRs (Beaumont et al. 2005).

Many pharmaceutical companies have responded energetically to these challenges and, driven by a better understanding of the underlying physiological processes, have moved away from traditional antagonist programs entering the quest to identify agonists, partial agonists, inverse agonists and allosteric modulators. The continuing advances in GPCR biology have provided more intervention points from agonist-dependent oligomerisation and signalling to the modification of responses by GPCR interacting proteins (for review, see Milligan 2004; Boeckart et al. 2004; May et al. 2004). Altering the deliverable from programmes and a focus on new modes of action has in many companies moved the emphasis from simple binding to functional in vitro assays for both lead identification and optimisation.

2 The Choice of Assay Technology

There is an ever-growing array of assay technologies and surrogate readouts to follow the activation of G-protein-coupled receptors (Fig. 1) and a detailed analysis is beyond the scope of this manuscript (for reviews, see Thomson et al. 2004, 2005; Williams 2004). The abundance of assay methods reflects the complex signalling mechanisms allowing multiple readouts, but also the fierce competition among providers of reagents and assay kits. As a consequence, there is some duplication and not all of the latest additions offer real scientific advantage or measurable benefit with respect to assay performance. This situation offers scientists the opportunity to choose methods fitting the local setting and instrumentation in their labs without compromising the biological relevance.

The focus of assays within HTS and lead optimisation has moved away from simple binding assays towards functional assays, driven by the desire to explore new mechanisms of action (Williams and Sewing 2005) and a stronger focus on agonists or allosteric modulators. The de-

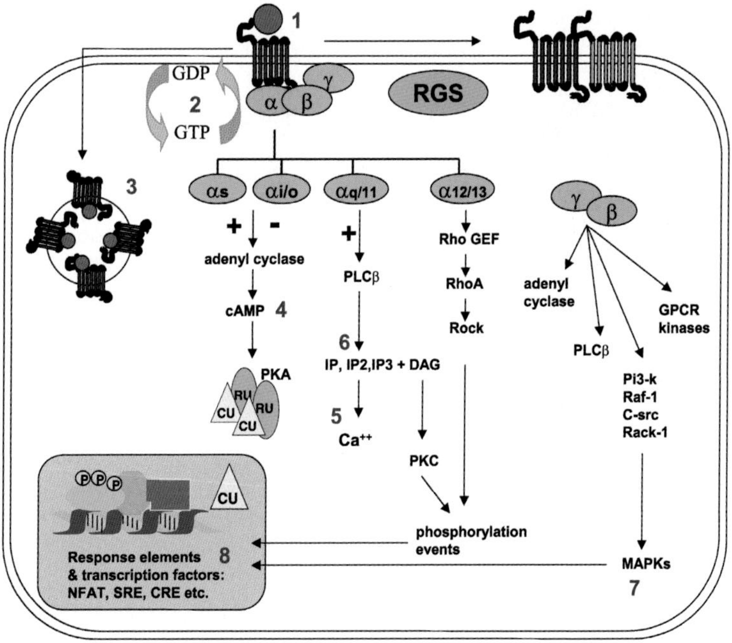

Fig. 1. Signalling through GPCRs. Overview of the multiple signalling events following activation of GPCRs and the key surrogate read-outs/assay technologies applied to measure receptor activation. *1* Receptor binding (fluorometric and radiolabelled ligands). *2* Measuring GDP/GTP exchange (GTPyS assays). *3* Receptor internalisation (translocation assays with fluorescent labelling of receptor, ligand or associated proteins). *4–6* Measuring second messengers (cAMP, calcium, IP3 etc.). *7* Map kinase activation (translocation of activated enzyme to the nucleus). *8* Reporter genes linked to response elements specific for the signalling under investigation. *RGS*, regulators of G-protein signalling

cision of where to measure compound activity is especially important for agonist programmes, as different levels of sensitivity and potency (Fig. 2) can be observed depending on the point of measurement. In addition, when considering antagonist programmes, the increasing number of GPCRs where allosteric modulation is described in the literature (May et al. 2004) presents a compelling case for using functional as-

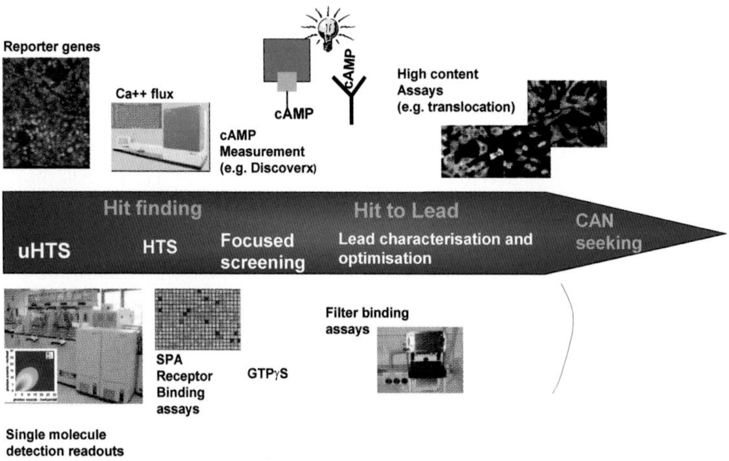

Fig. 2. HT assay technologies: from uHTS to CAN seeking. *HTS*, high-throughput screening; *uHTS,* ultra-HTS (100,000 compounds/day). For details, see text

say systems. As a rule of thumb, it should be noted that moving down the signalling cascade (using cell-based assays), increases the number of non-specific hits but will provide more sensitivity when looking for agonists and will potentially increase the assay window in antagonist settings.

The biochemical and cell-based assay technologies applied for GPCR signalling cascades can be broadly divided into five categories:

- Measuring receptor binding
- G-protein activation
- Detection of secondary messengers
- Reporter genes
- Protein interaction/translocation assays

Choosing assay technologies should start with an assessment of do-ability (equipment, expertise, and ready-made reagents) and through-put need: for example, a filter binding assay is still acceptable when used as an infrequent selectivity assay but is not suitable for supporting higher-throughput lead optimisation as primary assay (see also Fig. 3). Biological relevance and validation is key to success, but with a move toward targets without literature references, the validation with known compounds will not always be possible. In these cases, the assay development will focus on making the assay sensitive (i.e. working at K_d for ligand-binding assays, EC50 for functional antagonist assays, etc.). A sensitive assay is key, and this has often been achieved through the use of recombinant systems with highly over-expressed targets; however, the ongoing challenge is the relevance of these systems when compared to more relevant systems (i.e., cell lines endogenously expressing the target or ex vivo tissue preparations). The same question is true for the application of binding assays, as binding has to be translated into function and the relation/translation of different assays should be ideally known upfront or generated early in the project.

Cost will play a role in most organisations, but the cost has to be calculated beyond the assay group, as a more costly assay giving better quality results is more than balanced by the fact that whole teams of chemists work on the target and a saving in the screening team may be short-sighted.

As a detailed analysis is beyond the scope of this manuscript, two examples should highlight important developments during the last 5 years: within high-throughput screening, growing compound files have driven miniaturisation and, due to the limited potential to miniaturise radioligand binding assays, have driven the application of fluorescence readouts (for review see Jäger et al. 2003, Gribbon and Sewing 2003). Single-molecule spectroscopy, for example, one- and two-dimensional fluorescence distribution analysis, is ideally suited because these methods are, in theory, independent of the assay volume and provide a valuable tool for uHTS (Rüdiger et al. 2001). With the application of fluorescence readouts, compound interference due to auto fluorescence as well as light scattering (in fluorescence polarisation assays) has driven the research for more robust fluorescence methods. Apart from red-shifted fluorescence labels, time resolved methods, or fluorescence lifetime analysis, advanced data analysis methods based on the multiple parameters measured in single-molecule spectroscopy can detect, and in some cases correct, compound-related artefacts (Gribbon and Sewing 2003).

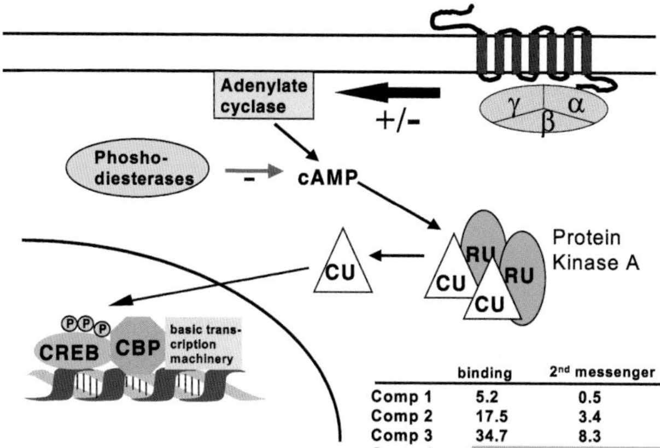

	binding	2nd messenger	reporter gene
Comp 1	5.2	0.5	0.11
Comp 2	17.5	3.4	0.9
Comp 3	34.7	8.3	2.7
Comp 4	82.9	12.3	3.2
Comp 5	89.1	11.5	3.5
Comp 6	175.2	15.3	8.3

all results in nM

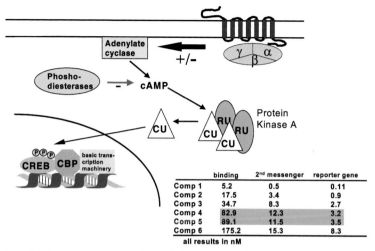

Fig. 3. Measuring agonist activity. Schematic and simplified representation of signalling through the activation of the adenylate cyclase pathway. Indicated in the table are the potencies of selected compounds applying different assay read-outs within an example project for illustration of the principle only. *Binding*: Whole cell binding with radiolabelled ligand. *2nd messenger*: Quantification of cAMP levels. *Reporter gene*: measuring receptor activation through a reporter gene couple to a cAMP-responsive element

Assays based on sub-cellular imaging and automated image analysis have provided a new tool in the area of functional cell based assays and are collectively classified as high-content screening (HCS). Initially based on immunofluorescence methods and fixed cells, the development of biochemical sensors provides the tools to follow signalling events in real time (for review, see Guliano et al. 2003). In HCS, multiple assay parameters can be analysed in parallel (cell shape, cell viability, translocation events, etc.), giving more information than traditional readouts. Biosensors are derived by labelling macromolecules with small fluorescence dyes or by applying genetic engineering to construct chimeras of cellular proteins fused to naturally fluorescent proteins, such as GFP, to establish a genetically encoded sensor. For example, the fusion of β-arrestin with GFP established a more generic procedure to measure activation of a number of different GPCRs through recruitment of GFP-arrestin to the plasma membrane (Barak 1997). The HCS assays (for review, see Milligan 2003; Grånäs et al. 2003) are widely commercialised, with companies like Cellomics and BioImage providing the analysis of whole signalling cascades with an array of readily available assays for compound profiling.

HSC assays are read on specially adapted imaging systems, loosely based around inverted fluorescence microscopes. The need for rapid and reliable automatic focusing, the quantity of image-derived data, and the complexity of algorithms for accurate image analysis have restricted the use of these assays to date. Driven by further technical development, the first examples of screening campaigns for lead identification have been described, but most companies apply HCS assays in later stages, mainly for compound series and lead characterisation.

3 Reagent Generation and Supply

High-quality reagents are key to screening campaigns with several million compounds, but also for the support of lengthy lead optimisation processes. For GPCR targets using functional, cell-based assays, the quantity and quality of biological screening has often been limited by the capacity and consistency of the cell supply. Many of the preconceptions relating to the use of live cells (Moore and Rees 2001; Williams and Sewing 2005) are based on the difficulties of maintaining a timely and high-quality supply of cells. The introduction of advanced, reliable cell culture automation has been a prerequisite to the application of functional cell-based assays in higher-throughput lead finding. Where adherent cells are used the introduction of systems such as SelecT (The Automation Partnership, Royston, UK) provide the basis for 24/7 cell supply for screening without human intervention and working beyond core working hours. This strategy has worked well for HTS, but the ongoing parallel support for dozens of lead optimisation projects has been resource-intensive, even with the application of automation, as programming and running of automated systems is ideally suited for large batches but less efficient for multiple small cell batches. The introduction of frozen cells (Zuck et al. 2004; Kunapoli et al. 2005) has transformed work with cell-based reagents as large batches of cells can be produced, quality controlled, and frozen and aliquots thawed and seeded on demand. This technique has provided a simple supply logistic (Fig. 4) and brought cell-based reagents in line with biochemical assays with respect to the ease of application.

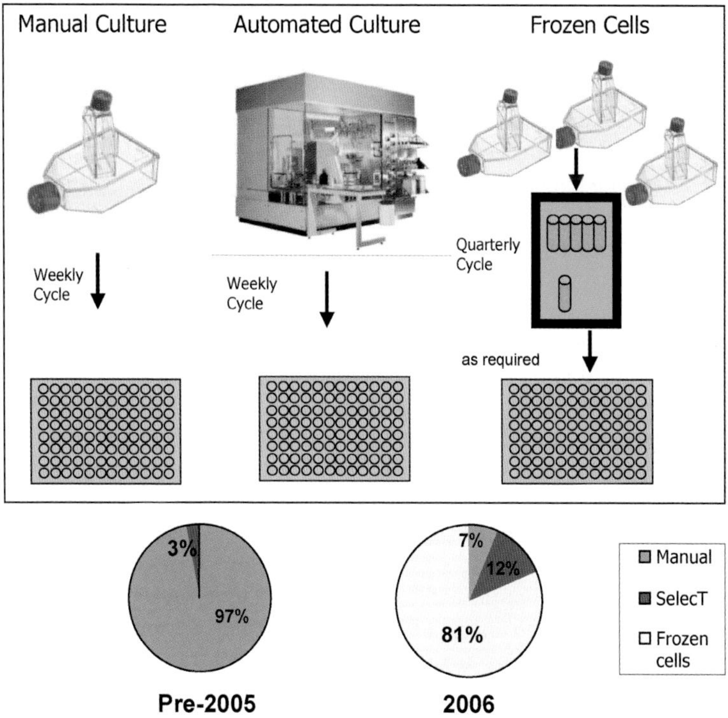

Fig. 4. Supporting lead optimisation for GPCR directed projects with live cells. Shown are the three methods applied for weekly project support and their changing, relative contributions to the overall project support from pre-2005 to 1Q 2006. Manual (maintenance and plating by cell culture scientist) and automated culture require tight scheduling and logistics, whereas the frozen cell approach can supply cell on short notice (on demand). For further details see text

Genetic engineering is a major element in the generation of cell-based assay reagents as endogenously expressing cell lines (due to limitations of supply, consistency, difficult handling, etc.) are rarely applied early in lead finding and optimisation, often restricted to scenarios where there is no freedom to operate. To speed up assay development, molecular toolboxes have been created (panels of ready-to-clone cell lines incorporating reporter genes with varying response elements). Although there is an increasing number of references describing transient expression systems for compound characterisation, larger campaigns still mainly rely on the generation of stable cell lines, and few examples are described where large-scale operations are supported by transient systems. Despite the complex biology and the discovery of cell type-specific accessory proteins modulating GPCR signalling, do-ability aspects (ease of handling, cost and the toolbox approach described earlier) have driven the use of two simple and robust cell lines, HEK293 and CHO K1, which remain the workhorses for HTS and higher-throughput lead optimisation work in most laboratories; and in our experience these two cell lines cover more then 80% of projects. Beyond mammalian cells, yeast (Minic et al. 2005) and frog melanocytes (Carrithers et al. 1999) are used for GPCR-directed screening, and platform technologies are commercially offered as a service from specialised providers (for example, Arena Pharmaceuticals, San Diego, CA, USA).

The generation of cell lines is standardised in most companies, routinely applying FACS for the generation of clones with different expression levels. A thorough characterisation of clones is needed to drive selection, as different stages of discovery have different needs with respect to reagents. For primary screening, sensitivity is key, whereas in lead optimisation the differentiation of compounds, for example, full vs partial agonist, can play a critical role and may not be possible with the same clone or assay technology.

4 Lead Finding Strategies Applying Biological Screening

For targeting GPCRs, biological screening remains an important tool in the lead finding strategy of most organisations. With the increasing availability of structural information and chemistry expertise, screening campaigns are complemented and sometimes replaced by knowledge-based approaches. Rather then being in competition, knowledge-based approaches allow a more informed choice as to which compounds should be screened, since a central question within the high-throughput screening area remains how many compounds have to be screened for success? Although company collections have grown to many millions, there is still a large gap between real collection size and theoretical calculation of how many compounds one would have to screen to find a lead for every target (Wintner and Moallemi 2000). In some companies, full file screening, i.e. screening all available compounds for all targets, is no longer the default strategy, as cost and timelines remain high not withstanding miniaturisation and uHTS. Iterative screening strategies are applied and can offer a cost and resource sparing strategy for some GPCR targets, for instance when searching for antagonists for aminergic GPCRs (due an abundance of chemical matter in many compound files relating to previous chemistry programmes). Iterative screening in its simplest disguise starts with screening part of the file and reviewing the identified chemical matter to decide on the need for more screening. More sophisticated approaches will cluster the compound file and screen varying numbers of signposts for the clusters rather than all compounds in the collection. Confirmed actives from screening are starting points for nearest neighbour searches in the compound file, followed by another round of screening, with a varying number of iterations to complete the approach.

Beyond the numbers game in primary screening, there have been general questions raised about the validity of the isolated target paradigm and molecular mechanism-based screening (Sams-Dodd 2005). The use of pathway or even whole organism screening has been presented as a new concept to identify compounds with new modes of action, but also to target multiple steps in signalling cascades simultaneously. Ultimately this leads to the concept of phenotypic screening where initially no prior target knowledge is required, but compounds are selected by the induced phenotype (for reviews, see Clemens 2004; Wakatsuki et al. 2004; Austen and Dohrmann 2005). However, this requires a general change in thinking in all areas of drug discovery and is challenging when driving SAR as well as safety studies.

5 Lead Selection

As discussed earlier, GPCR-directed drug discovery has been very successful and produced a number of blockbuster drugs. Looking at the analysis published by Klabunde and Hessler (2002) it becomes obvious that success is not evenly spread within the family, as two-thirds of the described examples are directed at GPCRs with bioamine ligands. Although a time bias may be inherent in this type of analysis, there is widespread agreement that there is a class difference and programmes aimed at GPCRs with peptide ligands have been more difficult to execute. The lower success rate of these projects has been attributed to the fact that drug molecules frequently mimic the physicochemistry of the receptor ligand (Beaumont et al. 2005). These characteristics are not necessarily compatible with small, orally delivered drug molecules (best described by the RoF paradigm; Lipinski et al. 1997), which remain at the centre of many drug discovery efforts. The concepts of target and ADME space as defined sub-regions of the overall chemical space are generally accepted today. ADME space is defined by the physicochemical properties of compounds being compatible with the requirement for oral absorption, dynamics, metabolism and excretion of compounds. The target space contains all molecules able to modulate a given target or family of targets such as GPCRs, kinases, etc. Further subdividing the GPCR family, it becomes clear that there is a different amount of overlay of sub-families of GPCR and the loosely defined ADME space. Indeed, aminergic GPCRs and GPCR with peptide ligands can be distinguished by the physicochemical profile of properties of populations of

→

Fig. 5a–c. Theory of ligand efficiency-based compound selection. Shown is the analysis and selection of active compounds from a larger set of compounds based on ligand efficiency and the impact on the molecular weight profile of the resulting compound sub-set to illustrate the general approach taken. **a** Ligand efficiency selection. Marked in *red* are compounds selected based on their high ligand efficiency (cut-off, 0.4) **b** Distribution of "actives based on ligand efficiency"; with respect to potency. **c** Molecular weight and clogP profile of the whole compound set. Marked in *red* are the compounds selected based on ligand efficiency

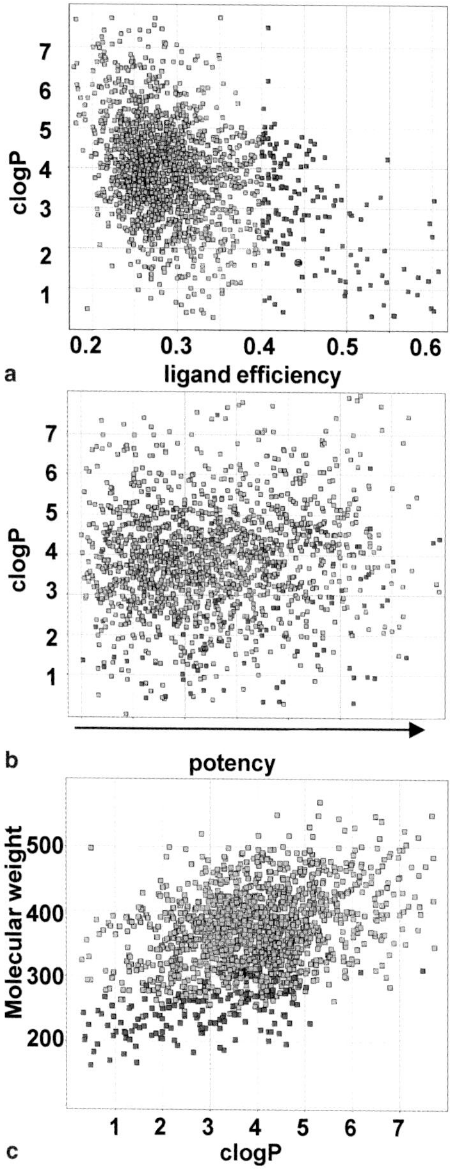
7
6
5
clogP 4
3
2
1
0.2 0.3 0.4 0.5 0.6
a
ligand efficiency
7
6
5
clogP 4
3
2
1
b
potency
Molecular weight
500
400
300
200
1 2 3 4 5 6 7
c
clogP

actives identified from biological screening (Nowlin et al. 2006). Conducting this type of analysis on a set of more than 27,000 confirmed actives derived from GPCR-directed screening campaigns reveals that the average RoF failure rate per compound is 0.276 in the case of aminergic GPCRs and a much higher failure rate of 0.418 per compound identified from screening against GPCRs with peptide ligands.

A focus on favourable physicochemical properties is already needed in the lead selection (and optimisation) strategy, and this has been recognised widely within the industry. For example, it has been noted that an early focus on potency biases compounds towards higher molecular weight (for example, Hann and Oprea 2004; Gribbon and Sewing 2005) and there have been attempts to normalise potency by taking the molecular weight, or more precisely the number of non-hydrogen atoms, into account. Initially based on the work of Kuntz (Kuntz et al. 1999), the concept of ligand efficiency-based compound selection has been introduced (Hopkins et al. 2004) and applied to hit and series selection (see Fig. 5). More complex calculations take multiple variables into account. They give the scientist a tool to establish a rank order for compounds for follow-up but also to monitor lead optimisation in multiple dimensions by reducing the complexity of data (Abad-Zapatero and Metz 2005).

6 Hit-to-Lead and Lead Optimisation

There has been a major focus on the hit-to-lead and lead optimisation process in many companies (Alanine et al. 2003; Davis et al. 2005), mainly driven by two factors: the increasing overall cost of drug discovery has put pressure on biology and chemistry teams to execute multiple projects in parallel with reduced resource time, and as a consequence there has been a need for better process definition and subsequent process improvement. The second driver is the new paradigm of multidimensional compound optimisation (Caldwell et al. 2001; Di and Kerns 2003), focusing in parallel on potency, selectivity and a panel of ADME parameters (as the in vivo activity of compounds is the result of structural elements and physicochemical properties) combined with early information on safety (mostly centred around HERG liabilities). This has increased workload in early discovery and exposed the interface between adjacent areas as key for seamless project execution (with respect to information flow). The goal is the efficient completion of each iteration in the lead optimisation process within a well-defined timeline, a concept that is termed "closed-loop" at Pfizer. Closed-loop describes the cycle (and interplay) through the main work areas of compound design, synthesis and purification, and compound flow and screening. Whereas previously decisions relied on anecdotal knowledge and the experienced medicinal chemist, the new process is data driven and focuses on "objective decisions" driven by software wherever possible. Only this "automation" will allow organisations to cope with the sheer volume of data from HTS hit lists, the parallel optimisation of multiple compound series, and the multiple values derived for each compound in lead optimisation.

Visualisation of multiple parameters is challenging for the scientist and there are developments to provide a reliable, more accessible, ranking of compounds. Ligand matrices have been introduced (Abad-Zapatero and Metz 2005) to reduce the number of variables and are a further development of the concept of ligand efficiency described earlier. Combining varying physicochemical property values (e.g. molecular weight, clogP, polar surface area, etc.) with potency, these methods aim to provide a robust basis for selection and to drive optimisation by introducing numerical rules. It is important to stress that these developments aim to help the medicinal chemist rather then delegating decisions to IT systems.

7 Conclusions

Lead finding for GPCR targets has shifted from simple binding assays to the use of functional assay systems previously only applied as second-line assays. The complex biology of GPCR mediated signalling and the potential multiple intervention points for drug therapy will continue to drive the development of complex assay systems with the prospect of targeting whole pathways or using phenotypic screens to move beyond the isolated target paradigm. Within lead selection and subsequent optimisation, a shift from purely potency-based selection towards evaluation of physicochemical properties predictive of in vivo behaviour has driven the introduction of parallel screening for potency, selectivity, safety and ADME properties. Together these developments provide the opportunity to break into new territory to reconcile the requirement of ADME with potency at the selected target driven by in vitro assays.

References

Abad-Zapatero C and Metz JT (2005) Ligand efficiency indices as guideposts for drug discovery. DDT 10:464–469

Alanine A, Nettekoven M, Roberts E, Thomas AW (2003) Lead generation – enhancing the success of drug discovery by investing in the hit to lead process. Comb Chem High Through Screen 6:51–66

Austen M, Dohrmann C (2005) Phenotype-first screening for the identification of novel drug targets. DDT 10:275–282

Barak LS, Ferguson SSG, Zhang J, Caron MG (1997) A beta-arrestin green fluorescent protein biosensor for detecting G protein-coupled receptor activation. J Biol Chem 272:27497–27500

Beaumont K, Schmid E and Smith DA (2005) Oral delivery of G protein-coupled receptor modulators: an explanation for the observed class difference. Bioorg Med Chem Lett 15:3658–3664

Bockaert J, Fagni L, Dumuis A, Marin P (2004) GPCR-interacting proteins (GIP). Pharmacol Ther 103:203–221

Caldwell GW, Ritchie DM, Masucci JA, Hageman W, Zhengyin Y (2001) The New Pre-preclinical paradigm: compound optimisation in early and late phase drug discovery. Curr Top Med Chem 1:353–366

Carrithers MD, Marotti LA, Yoshimura A, Lerner MR (1999) A melanophore–based screening assay for Erythropoietin receptors. J Biomol Screen 4:9–14

Clemons PA (2004) Complex phenotypic assays in high-throughput screening. Curr Opin Chem Biol 8:334–338

Davis AM, Keeling DJ, Steele J, Tomkinson NP, Tinker AC (2005) Components of successful lead generation. Curr Topic Med Chem 5:421–439

Di L, Kerns EH (2003) Profiling dug-like properties in discovery research. Curr Opin Chem Biol 7:402–408

Fredriksson R, Schiöth HB (2005) The repertoire of G-protein-coupled receptors in fully sequenced genomes. Mol Pharm 67:1414–1425

Glasel JA (2004) Emerging concepts in GPCR research and their implications for drug discovery. Decision Res www.decisionresources.com. Cited 26 November 2006

Grånäs C, Lundholt BK, Heydorn A, Linde V, Pedersen HC, Krog-Jensen C, Rosenkilde MM, Pagliaro L (2005) High content screening for G protein-coupled receptors using cell-based protein translocation assays. Comb Chem High Throughput Screen 8:301–319

Gribbon P, Sewing A (2003) Fluorescence readouts in high throughput screening: no gain without pain? DDT 8:1035–1043

Gribbon P, Sewing A (2005) High-throughput drug discovery: what can we expect from HTS? DDT 10:17–22

Giuliano KA, Haskins JR, Taylor DL (2003) Advances in high content screening for drug discovery. Assay Drug Dev Technol 1:565–577

Hann MM, Oprea TI (2004) Pursuing the lead likeness concept in pharmaceutical research. Curr Opin Chem Biol 8:255–263

Hefti E, Bolten BM (2000) Advances in high throughput screening – do they lead to new drugs? Decision Resources, October 2003, www.decisionresources.com. Cited 26 November 2006

Hopkins AL, Groom CR (2002) The druggable genome. Nat Rev Drug Disc 1:727–730

Hopkins AL, Groom CR, Alex A (2004) Ligand efficiency: a useful metric for lead selection. DDT 9:430–431

Jäger S, Brand L, Eggeling C (2003) New fluorescence techniques for high-throughput drug discovery. Curr Pharm Biotechnol 4:463–476

Klabunde T, Hessler G (2002) Drug design strategies for targeting G-protein-coupled receptors. Chem Bio Chem 3:928–944

Klumpp M, Scheel, Lopez-Calle, Busch, Murray J, Pope AJ (2001) Ligand binding to transmembrane receptors on intact cells or membrane vesicles measured in a homogeneous 1-microliter assay format. J Biomol Screen 6:159–170

Kunapili P, Zheng W, Weber M, Solly K, Mull R, Platchek M, Cong M, Zhong Z, Strulovici B (2005) Application of division arrest technology to cell-based HTS: comparison with frozen and fresh cells. Assay Drug Dev Techn 3:17–26

Kuntz ID, Chen K, Sharp KA, Kollman PA (1999) The maximal affinity of ligands. Proc Acad Nat Sci USA 96:9997–10002

Lipinski C, Lombardo F, Dominy BW, Feeney PJ (1997) Experimental and computational approaches to estimate solubility and permeability in drug discovery and development settings. Adv Drug Deliv Rev 23:3–25

May LT, Avlani VA, Sexton PM, Christopulus A (2004) Allosteric modulation of G protein-coupled receptors. Curr Pharm Des 10:2003–2013

Milligan G (2003) High-content assays for ligand regulation of G-protein coupled receptors. DDT 8:579–585

Milligan G (2004) G protein-coupled receptor dimerisation: function and ligand pharmacology. Mol Pharmacol 66:1–7

Minic J, Sautel M, Salesse R, Pajot-Augy E (2005) Yeast system as a screening tool for pharmaceutical assessment of G protein-coupled receptors. Curr Med Chem 12:961–969

Moore K, Rees S (2001) Cell-based versus isolated target screening: how lucky do you feel? J Biomol Screen 6:69–74

Nambi P, Aiyar N (2003) G protein-coupled receptors in drug discovery. Ass Drug Dev Tech 1:305–310

Nowlin D, Bingham P, Berridge A, Gribbon P, Laflin P, Sewing A (2006) Analysing the output from primary screening. Com Chem High Trough Screen 9:331–337

Peakman T, Franks S, White C, Beggs M (2003) Delivering the power of discovery in large pharmaceutical organisations. DDT 8:203–211

Presland J (2005) Identifying novel modulators of G protein-coupled receptors via interaction at allosteric sites. Curr Opin Drug Disc Dev 8:567–576

Rüdiger M, Haupts U, Moore KJ, Pope AJ (2001) Single-molecule detection technologies in miniaturised high throughput screening: binding assays for G protein-coupled receptors using fluorescence intensity distribution analysis and fluorescence aninsotropy. J Biomol Screen 6:29–37

Sams-Dodd F (2005) Target-based drug discovery: Is something wrong? DDT 10:139–147

Thomsen WJ, Gatlin J, Unett DJ, Behan DP (2004) Developing functional GPCR screens. Curr Drug Disc 1:13–18

Thomsen W, Frazer J, Unett D (2005) Functional assays for screening GPCR targets. Curr Opin Biotech 16:655–665

Wakatsuki T, Fee JA, Elson EL (2004) Phenotypic screening for pharmaceuticals using tissue constructs. Curr Pharm Biotech 5:181–189

Williams C (2004) cAMP detection methods in HTS: selecting the best from the rest. Nat Rev Drug Discov 3:125–135

Williams C, Sewing A (2005) G-protein-coupled receptor assays: to measure affinity or efficacy, that is the question. Comb Chem High Through Screen 8:285–292

Wintner EA, Moallemi CC (2000) Quantized surface complementarity diversity (QSCD): a model based small molecule-target complementarity. J Med Chem 43:1993–2006

Young SS, Lam RLH, Welch WJ (2002) Initial compound selection for sequential screening. Curr Opin Drug Disc Dev 5:422–427

Zuck P, Murray EM, Stec E, Grobler JA, Simon AJ, Strulovici B, Inglese J, Flores OA, Ferrer M (2004) A cell-based b-lactamase reporter gene assay for the identification of inhibitors of hepatitis C virus replication. Ana Biochem 334:344–355

Ernst Schering Foundation Symposium Proceedings

Editors: Günter Stock
Monika Lessl

Vol. 1 (2006/1): Tissue-Specific Estrogen Action
Editors: K.S. Korach, T. Wintermantel

Vol. 2 (2006/2): GPCRs: From Deorphanization to Lead Structure Identification
Editors: H.R. Bourne, R. Horuk, J. Kuhnke, H. Michel

Printing: Krips bv, Meppel
Binding: Stürtz, Würzburg